Can Reindeer Fly?

THE SCIENCE OF CHRISTMAS

Can Reindeer Fly?

THE SCIENCE OF
CHRISTMAS

Roger Highfield

Weidenfeld and Nicolson
LONDON

First published in Great Britain in 1998 by
Metro Books (an imprint of Metro Publishing Limited),
19 Gerrard Street, London W1V 7LA

This revised edition published in 2001
by Weidenfeld & Nicolson

The author is grateful for permission to include the
following previously copyrighted material: 'Talking Turkeys',
copyright © Benjamin Zephaniah, 1994. Taken from the book
Talking Turkeys, published by Penguin Books.

A CIP catalogue reference for this book
is available from the British Library

ISBN 0 297 60774 X

Typeset by Selwood Systems, Midsomer Norton

Printed in Great Britain by
Butler & Tanner Ltd, Frome and London

Weidenfeld & Nicolson

The Orion Publishing Group Ltd
Orion House
5 Upper Saint Martin's Lane
London, WC2H 9EA

Contents

In memory of my father

Acknowledgements

I have endeavoured in this Ghostly little book, to raise the Ghost of an Idea, which shall not put my readers out of humour with themselves, with each other, with the season, or with me. May it haunt their house pleasantly, and no one wish to lay it.

Their faithful friend and servant, C.D.
December 1843

Like Charles Dickens in his preface to *A Christmas Carol*, I propose 'to raise the Ghost of an Idea'.

In Dickens's classic book, the spirits of Christmas Past, Present and Future reveal the true meaning and spirit of the season to Ebenezer Scrooge, transforming him from a miser into a potent symbol of charity.

I, too, hope to enlighten the reader by acting as a guiding spirit, one who will illuminate Christmas by viewing the holiday and its rituals from a new perspective, that of science. Christmas and associated celebrations offer a wonderful excuse to explore a broad range of fields, from biotechnology and fractals to neuropharmacology and nanotechnology. If appetites are whetted for science, or at the very least curiosity about the subject is stimulated, I will be pleased. Any change in the charitable behaviour of the reader would, of course, be a welcome bonus.

Each Christmas for more than a decade I have written about seasonal science for the *Daily Telegraph*. Many thanks to my editor, Charles Moore, and his predecessor, Max Hastings, for indulging my obsession. Gulshan Chunara, as ever, provided me with invaluable assistance. It has also been stimulating discussing aspects of the book with my colleagues, Adrian Berry, David Derbyshire, Aisling Irwin, David Johnson, Laura Spinney, Tom Standage and Robert Uhlig.

John Brockman and Katinka Matson provided the encouragement to first develop a proposal. Very many thanks are also due to Little, Brown and Company, Rowohlt and Weidenfeld & Nicolson for backing the project and in particular to Rick Kot for his warm encouragement and support.

I would like to thank the following people and organizations for helping me to seek out the science behind the festivities: Leonard Adleman, Denis Alexander, Anthony Astbury, Linda Bartoshuk, Gerard Bond, the British Antarctic Survey, Donald Brownlee, L.P. Bucklin, Stephen Burley, David Cheal, Chris Clayton, Malcolm Cooper, Winnifred Cutler, Piero Dolara, Jonathan Dorfman, Len Fisher, Jeffrey Friedman, Takanari Gotoda, Steven Guest, Alan Hirsch, Nina James, Charles Jenkins of NASA, Steve Jones, Barry Kemp, Krishna Podila, Sir Harry Kroto, Laurie Lucchina, Patrick McGovern, George Masterton, Dale Matthews, Wendy Mechaber, John Metz, Michael Molnar, John Moore, Kenneth Pargament, David Peel, Raj Persaud, David Phillips, Bill Proebsting, Sir Martin Rees, Norman Rosenthal, Nigel Scott, David Skuse, John Maynard Smith, Kristina Staley, Andrew Strassman, Joergen Taageholt, Fred Turek, UNICEF, Alan Watkins, George Williams, Ian Wilmut, Lewis Wolpert and Xiao Zeng.

Ronald Parkinson of the Victoria and Albert Museum in London was kind enough to spend a morning with me discussing the museum's vast collection of Christmas cards. Linda Capper also proved a great help when it came to contacting members of the British Antarctic Survey.

A number of people have also read parts or all of the manuscript to ensure that the science is understandable. Many thanks to: my wonderful wife, Julia; my daughter Holly; my parents,

Acknowledgements

Ron and Doris Highfield; and to a number of friends: Samira Ahmed, Peter Coveney, Tony Manzi, Eamonn Matthews, Brian Millar, Sharon Richmond and Martin Winn. The Revd Dr John Platt of Pembroke College, Oxford, should be mentioned for his kind reassurance that the religious sections of my book are not blasphemous.

Particular thanks to Graham Farmelo of the Science Museum, for his many and constructive suggestions on an early draft, to Robert Matthews for his calculating skills and his Murphy's Law expertise, and to my mother for translating German language papers and Christmas books.

I am also indebted to a number of researchers who gave me feedback on specific sections of the book. I have covered such a wide range of fields that I am confident of one thing: a number of howlers remain, all of which are my responsibility. Many thanks to the following, for helping me weed out some of the worst: Miguel Alcubierre, Peter Atkins, Peter Barham, Charles Bennett, Sam Berry, David Bonthron, Roy Bradshaw, Samuel Braunstein, Roger Buckland, Carole Burgoyne, Linda Capper, Isaac Chuang, David Clary, Angela Clow, Roger Cone, Cary Cooper, Peter Coveney, Glenn Cox, Peter Davies, Leslie Dawes, Dan Dietrich, Robert East, Sabine Eber, Ron Evans, Len Fisher, Matthew Freeman, Adrian Furnham, David Gems, Alexei Glebov, Richard Gross, Rose Gubitosi-Klug, Sunil Gupta, Laurance Hall, Odd Halvorsen, Patrick Harding, Mike Hayden, Marion Hetherington, Paul Hoffman, James Horne, David Hughes, Ilpo Huhtaniemi, Colin Humphreys, Dan Keathley, David Kelly, Gerd Kempermann, Harold Koenig, Sandy Knapp, Tom Lachlan-Cope, Michel Laroche, Dale Lewison, Robin Lovell-Badge, Patrick McElduff, Stanley McKnight, Neil Martin, Dave Mela, Randolf Menzel, Daniel Miller, Les Noble, Adrian North, José Pardo, Daniele Piomelli, Caroline Pond, David Price-Williams, Wolf Reik, Allen Riordan, Margaret Robins, Delwen Samuel, Larry Silverberg, Gene Stanley, Ian Stewart, Scott Swartzwelder, Luca Turin,

Mark Uncles, Dietmar Voelkle, Nigel Weatherill, Bernard Wentworth, Diederik Wiersma, Andy Yeatman, Anton Zeilinger and Timothy Zwier.

God bless us, every one!

introduction
Christmas and
the Scientist

There seems a magic
in the very name of Christmas.
Charles Dickens, *Sketches by Boz*

What is your image of Christmas? Mulled wine, tinselled trees
and the crunch of snow? Or carols, family gatherings and gaudy
greetings cards? Science is probably the last thing you associate
with the festivities and yet Christmas is also a time to celebrate
research across a vast range of scientific disciplines.

Chemists are hard at work in the Christmas kitchen. Experts
on thermodynamics have drafted equations that help us cook
turkeys to perfection; pharmacologists have traced the metabolic pathways of the brain to explain why chocolates can be so
addictive; and steaming plum puddings have been scrutinized
by one version of the scanners used by doctors to peer inside
patients.

Meteorologists have studied every aspect of the snow cycle,
from the formation of an ice crystal high in the sky to the traces
of past Christmases buried deep in the snowpack. Climatologists
are using these data to help predict white Christmases far into
the future. A handful are even concocting outlandish schemes
to guarantee that each and every Christmas is white.

Psychologists tease out the hidden agenda of the Christmas
card – or present – and what it reveals about our social status.
All the while, anthropologists hunt for the meaning of the

1

celebrations in pagan rituals that took place long before the birth of Christ, during long winter nights when our ancestors feared the sun would never return.

The origins of the celebrations in the darkness of prehistory emphasize perhaps the most fundamental aspect of Christmas: everyone's invited. Today, the traditional Christmas hoopla takes place alongside both the Jewish celebration of Hanukkah and an African-American harvest holiday, Kwanzaa. The seasonal message of hope and charity is a message for all – Christians, Jews, Hindus, Moslems, Buddhists and yes, even scientists and engineers.

I have been investigating the science of Christmas for more than a decade. When I started to become interested in the subject, I was unprepared for the range of insights that would eventually emerge. Take, for example, those flying reindeer, Santa's red and white colour scheme and his jolly disposition. Far from being whimsical creations, they all have a scientific basis, being linked to the use of a hallucinogenic toadstool in ancient rituals. I can add that Santa was born with a genetic propensity to become obese and now suffers from diabetes. He does not live at the North Pole, preferring the warmth of an island off the coast of Turkey. There, panting at Santa's side, you will probably find Rosie, not Rudolph. (Rudolph is traditionally depicted adorned with antlers, but males lose their crowning glory around the time of the festivities.)

I was at first puzzled by how Santa could fly in any weather, circle the globe on Christmas Eve, carry millions and millions of presents and make all those rooftop landings with pinpoint accuracy. The answer lies in his unprecedented research resources and expertise across a range of fields, spanning genetic engineering, computing, nanotechnology, quantum electronics, and quantum gravity (see Glossary). My experience of writing this book undermines the idea that the materialist insights of science destroy our capacity to wonder, leaving the world a more boring and predictable place. For me, the very reverse is true. I can still remember the day when I first became convinced that Santa did not exist. Now, when the Santa myth is refracted through the prism of science, he seems more real than ever. I believe that science and technology can even shed a little light on a deeper question: where did Christmas come from in the first place? Peel back the wallpaper of centuries and you will find that the festival is an amalgam of a wide range of influences – German, Dutch, English, American and other traditions, both religious and pagan – that emerged over the millennia.

Part of the reason these ancient winter festivities went global can be found more than 150 years ago, at the tail end of the Industrial Revolution. it was then that 'Christ's Mass' (*Cristes Maesse* in Old English), the church service that celebrates the birth of Jesus Christ, along with a wealth of other traditions, entered the scientific age of mass communications, transport and other technologies. This collision between ancient tradition and the age of science and technology was particularly significant in the Victorian Britain of the 1840s. This decade alone saw a dizzying rate of change in society due to innovations across many different disciplines. In the world of science, these included Darwin's ideas on natural selection, Joule's work on thermodynamics and Faraday's studies of magnetism, light and electricity. In the sister disciplines of engineering and technology, Babbage was hard at work on his Difference Engine, a calculating machine, while information technologists were building a web of telegraph lines across the nation. The old certainties seemed to have been squashed flat by the steam hammer, steamboat and steam train. Breakthroughs in communication technology – from steam railways to the telegraph – allowed the old religious and pagan customs to be disseminated and homogenized for mass consumption. These have formed much of what we think of today as the 'traditional' festivities.

The tumultuous 1840s also saw the emergence of science as a discipline in its own right. William Whewell, a polymath who was a Fellow of the Royal Society, coined the word 'scientist' in his two-volume book *The Philosophy of the Inductive Sciences*. A hybrid of Latin and Greek, the word was attacked (wrongly) as 'an American barbarous trisyllable'. But the pressure to put a name to this increasingly influential group of innovators was overwhelming.

That same decade saw the introduction in Britain of one component of the German Christmas that remains very much a part of the celebrations today. In 1840 Queen Victoria and Prince Albert set up a Christmas tree for the first time in Windsor Castle. The Queen recorded that this German custom 'quite affected dear Albert, who turned pale and had tears in his eyes'. Eight years later Victoria and Albert appeared beside the tree in the *Illustrated London News*, giving rise to one of the most famous Christmas customs.

At the same time as the term 'scientist' was introduced, and Albert gazed upon his tree, another Christmas tradition was set in place by Henry Cole, an extraordinary individual whose achievements ranged from founding the Victoria and Albert Museum to organizing the Great Exhibition. He decided to reduce the burden of writing Christmas greetings letters by marrying mass communication with art. Cole's invention, the Christmas card, took advantage of another development in which he had a hand – the penny post. The first card was printed in 1843 and priced at one shilling, the equivalent of a day's wages for a labourer. After two decades the price fell dramatically thanks to one of the many technological innovations of the period, cheap colour lithography, and Christmas cards entered the mass market.

Henry Cole saw the card as the folk art of the Industrial Revolution and it became the greatest popularizer of Christmas iconography, with designs ranging from bizarre characters with pudding heads to mannequins in period costume as well as the more conventional mistletoe, robins, holly and fireside scenes. The cards were not simply printed on paper but were gilded, frosted, and dressed with satin or fringed silk. Some were even made to squeak.

Through the depiction of one of the card's most familiar characters, the fat descendant of the fourth-century St Nicholas, it is possible to trace the influence on our way of life of scientists, engineers and technologists in the wake of the pioneering contributions of Henry Cole, Prince Albert and William Whewell. The character I am referring to is, of course, that man with the white beard.

A silk-fringed card published in 1888 revealed how, by this time, Santa had resorted to the latest communications technology to improve links with his market. The figure shown on the card seems to be engaged in what can only be described as a conference call, listening to the simultaneous demands for presents from an assortment of children. Only a decade earlier, Alexander Graham Bell had patented the telephone that made it all possible. By the 1890s, Santa was depicted without his sleigh and reindeer, preferring to haul his gifts around by 'the new monstrosity from France', the automobile. Another newfangled device – the wireless – is seen on a Christmas card of 1929. Santa appears to be mesmerized by the crackling message being received over the ether: 'You're in my Christmas circuit/And on the waves of thought/A Happy Christmas and New Year/To you is gladly brought.' Radio would become the first mass medium to reinforce the tendency for Christmas to be a festival held behind closed doors. When Santa reached for a cool soda pop, in a Coca-cola advertisement from Christmas 1937, he was again a technological pioneer. The source of his refreshment was a refrigerator, at a time when ice boxes were still being used by most American households. Santa can now be found in cyberspace. Every Christmas, digital images of his fat frame scud about the web of international computer networks, spouting digital ho-hos and seasonal greetings. Henry Cole would be amazed by the extent to which his invention has caught on today. The significance of the 1840s does not end there, however. As Cole sent out his first cards, the greatest and most influential of all Christmas books appeared in a crimson and gold binding.

Charles Dickens's *A Christmas Carol* was published by Chapman and Hall on 19 December 1843. By Christmas Eve it had sold 6000 copies, the most successful publication that season. Within two months, eight pirated theatrical productions had been staged. The book had been inspired by Dickens's correspondence with the philanthropist Lord Ashley. Dickens was horrified by the impact of the machine age on society, particularly the appalling conditions endured by children working in coal mines and factories. He started work on his famous Christmas story to make what he called a 'Sledge hammer' blow against these evils of the industrial age. One newspaper described the *Carol* as 'sublime', while the novelist Thackeray called it 'a national benefit'. Lord Jeffrey told Dickens that it had 'prompted more positive acts of beneficence than can be traced to all the pulpits and confessionals in Christendom'.

Thus the 1840s saw a striking convergence – the first scientist, the first tree, the first card *and* the Christmas book to top all Christmas books. A century and half later, science is still altering the very nature and fabric of the celebrations through the introduction of new technology, whether in the form of cloned Christmas trees, electronic greetings on the Internet or those infuriating cards that play carols over and over again.

And so to the science of Christmas...

Roger Highfield
Greenwich, Christmas 1997, revised 2001.

one

The Bethlehem Star

Silent night! Holy night!
Guiding star, lend thy light!
J. Moier, 'Silent Night'

Two thousand years after it was first seen by the Wise Men, astronomers are still arguing about the Bethlehem Star. They suggest many explanations for this herald of the birth of Christ: a comet, the birth or death of a star, a conjunction of planets, an apparent hesitation in a planetary orbit, or even the sighting of the then-unknown planet Uranus.

One little-known fact is that the star was probably not the brilliant object portrayed on Christmas cards: it appears that King Herod and all his 'chief priests and scribes' missed it. St Matthew did not use the adjective 'bright' to describe it in his Gospel. Only in the early, less reliable, Christian literature does the star dazzle. In the Protoevangelium of St James, omitted when the Bible was compiled, the Wise Men declare: 'We saw how an indescribably great star shone among those stars and dimmed them.'

Heavenly objects did not have to be brilliant for the Wise Men to find them fascinating. The Magi attached a significance to cosmic events and structures that is quite alien to the thinking of their modern counterparts. Their perspective is highlighted by the translation of the Greek word *Magi*. The Authorized Version reads this as 'wise men' but the New English bible opts for 'astrologers'. Like good anthropologists, we must try to see the heavens through ancient eyes and minds to understand why this star was so significant in the Magi's Babylonian society.

The star makes one of its rare Biblical appearances in the Gospel according to St Matthew 2:1–12, which states that 'In the time of King Herod, after Jesus was born in Bethlehem of Judea, wise men from the East came to Jerusalem, asking, "Where is the child who has been born king of the Jews? For we observed his star at its rising, and have come to pay him homage."' Some theologians dismiss this reference to the star as a story made up to satisfy the Old Testament prophecy that 'A star shall come forth out of Jacob and a sceptre shall rise out of Israel.' Fulfilment of such a prediction would have provided succour for the faith. It certainly would have merited a mention, the biblical equivalent of 'I told you so!' Matthew's Gospel is full of references to the Old Testament, yet there is no such 'fulfilment statement' regarding the star. If we conclude, then, that this heavenly apparition was real, rather than something cooked up to satisfy an Old Testament prediction, what did the Wise Men see?

To identify the star of Bethlehem, all we have to do is assemble the clues in the Bible, work out what the heavens looked like around the time of the birth, and search for star candidates. Easy. If only it really were this simple, comments Bethlehem star theorist Colin Humphreys of Cambridge University: 'The difficulties of treating the star of Bethlehem as a real astronomical object should not be underestimated.' According to another star chaser, David Hughes, Reader in Astronomy at Sheffield University, drawing firm conclusions from the handful of mentions in the Bible is not easy: 'When you come to interpreting the facts, you have to pick and choose. You can't take them all on board.' One example of what David Hughes means is the description of the star's movements in Matthew 2:9–10: 'There, ahead of them, went the star that they had seen at its rising, until it stopped over the place where the child was.'

8

This kind of evidence stirs dissent. According to Hughes, 'Astronomical objects are a huge distance away and do not wander in front of people and then stand over and point out a specific house in a small village like Bethlehem.' Humphreys, however, believes this Biblical reference hints at the star's real identity: 'I spent quite a long time scouring ancient historical and astronomical literature and found two other references in which a star was said to "stand over" a place, both of which are comets.' Comets are chunks of frozen matter that sweep through elongated orbits around the Sun, warming to form a luminous tail of charged particles on their inward-bound trip.

Perhaps the star never existed. Did Matthew invent it to embellish the Nativity story? If so, he certainly did not describe it in the same exaggerated way as James, author of the Apocrypha gospel, whom tradition describes as the brother of Jesus. A similar reference is found in the letters of Ignatius, the first-century bishop of Antioch, who writes of a star that 'outshone all the celestial lights, and to which the Sun and Moon did obeisance'. I am inclined to believe, as Hughes puts it, that 'Matthew was a straight guy, telling it how it was'. Alas, Matthew is the only Gospel writer to mention the star. Luke's Nativity account tells of shepherds and their flocks, but does not mention the Magi, let alone the stellar subject of their quest. Mark and John only introduce Jesus as an adult.

WHEN WAS JESUS BORN?

Interpreting the meagre star evidence is tricky. When it comes to astronomy two millennia ago, there was no physical perspective and no astrophysics: the idea that planets differed from stars had not occurred to people. Instead, they were concerned with the relative position and motion of these points of light. Identifying the star would also be easier if we knew when Jesus was born. Then we could use a computer program to extrapolate from what we can see of the heavens today to what the Wise Men saw of them on that historic night. However, we don't have a precise date for Christ's birth.

Jesus was not born in the year AD 1, despite the fact that 'A.D.' stands for *Anno Domini* ('in the year of our Lord'). Our present calendar is a modification of the one introduced by Julius Caesar on 1 January 45 BC. The Roman system of dating *ab urbe condita* (from the foundation of Rome) began in the first century BC and lasted until the sixth century, when Denys (Dionysius Exiguus), a monk living in Rome, proposed that the Christian era should date from a unique event of far-reaching religious significance, the supposed year of Christ's birth. His system marked the origin of the AD sequence we now employ. Unfortunately, in reorganizing the calendar he overlooked four years of the rule of Augustus, suggesting that Christ was born around 4 BC. Further evidence that Denys was a menace arose because he lacked understanding of the concept of the number zero and labelled the first year AD 1. That meant one year had elapsed on December 31, of AD 1 and that the second millennium came to an end on December 31, AD 2000, not December 31, AD 1999, as most people seemed to think.

To pinpoint the year and date of Jesus's birth, one could try working back from Christ's Crucifixion. The Gospels state that this took place during the rule of Pontius Pilate as governor of Judea, which lasted from AD 26 to AD 36, according to the celebrated Jewish historian and Pharisee, Flavius Josephus. Colin Humphreys has determined the date that Jesus died on the cross as Friday 3 April AD 33, based on the date of the blood-red lunar eclipse that biblical and other references suggest followed the Crucifixion. But we don't know exactly how old Jesus was when he died: Luke says he was 'about thirty' when he started his ministry, while in another biblical reference, Jesus is told: 'You are not yet fifty.'

The Bible provides other clues to the birthday. Jesus was

born during the reign of Caesar Augustus, which narrows our search to some time between 44 BC and AD 14. Another clue is given by Matthew and Luke, who agree that Jesus was born during the reign of King Herod. It is generally accepted that Herod the Great died in the spring of 4 BC, though other dates have been put forward (5 BC, 1 BC and AD 1). He was replaced by his son Herod Antipas (21 BC–AD 39), who ruled throughout the ministry of Jesus. That also shrinks the timespan within which the birth took place but nowhere near enough to reveal what exactly the Wise Men were chasing.

Matthew 2:16 offers yet another hint for the star detective to chew over: 'Herod … killed all the children in and around Bethlehem who were two years old or under, according to the time that he had learned from the Wise Men.' So Jesus was probably born at least two years before the end of Herod's reign.

Another indication of the birth year comes from a reference to a Roman census that induced Joseph and Mary (who was 'great with child') to travel to Bethlehem (Luke 2:1–7): 'In those days a decree went out from Emperor Augustus that all the world should be registered. This was the first registration, taken while Quirinius was governor of Syria.' But to interpret this apparently simple statement is to enter a minefield of contradictions. There is no official record of a census by Publius Sulpicius Quirinius, who became governor in AD 6. He did conduct a census in AD 6–7, but that was of Judea, not of Galilee. There is a reference in Luke 2:1–5 to a census by Emperor Caesar Augustus around the time of the birth of Jesus. Records reveal three well-documented censuses conducted for Augustus, in 28 BC, 8 BC and AD 14, but involving only Roman citizens. These seem wide of the mark but Colin Humphreys points out that the fifth-century historian Orosius and the Jewish historian Josephus do refer to a census of allegiance to Augustus at the time of the birth of Christ. Perhaps this is what Luke meant. Confused? A great deal has been written about what census the New Testament is referring to. There is no need to go into any more detail here save to say that, after sifting the available evidence, many scholars estimate that Jesus was born some time between 4 and 7 BC.

This timescale rules out some candidate 'stars', such as Halley's Comet in 12 BC, or the conjunctions of Venus and Jupiter on 12 August 3 BC and 17 June 2 BC. The latter is a pity because on 17 May 2000, the planets merged again, suggesting a possible Second Coming – though the encounter was too close to the sun to see with the naked eye and was only visible to the Soho space probe.

We can use similar detective work to narrow down the time of year of Christ's birth. A date of 25 December is unlikely. Spring, or even autumn, seems a better bet, in the light of another icon of the Nativity, referred to by Luke: 'There were in the same country shepherds abiding in the field, keeping watch over their flocks by night.' The shepherds were most likely to be with their flocks during the lambing season, in the spring, or in the autumn, when the flocks were being collected. Indeed, some Christians have already celebrated 17 April 1995 as Jesus's 2,000th birthday. Others, such as David Hughes, stick to autumn: 'This is because the birth of John the Baptist was thought to have been in late March, and Jesus, his cousin, was six months younger.'

BETHLEHEM STARS

Working on the assumption that the period in which the birth of Jesus took place is known – between 4 and 7 BC, some time around September or March – we can now draw up a short list of candidates for the Bethlehem star. As long ago as AD 248, Origen (Origenes Adamantius), the celebrated Christian writer, teacher and theologian, suggested that the Bethlehem star was a comet. Perhaps it was the 'broom star' (*sui-hsing*) – so called because the comet's tail appeared to be sweeping the sky – that

was described in 5 BC by Chinese astronomers and was recorded in the official history of the Han dynasty. Colin Humphreys
argues that this was the event that induced the Magi to under-
take their journey to Jerusalem. If we accept this star candidate,
then the first Christmas was in the spring of 5 BC. We can
narrow down the season because the Chinese astronomers
recorded that the 'broom-star' comet appeared in *Ch'ien-niu*
which, according to ancient star maps, is the area of the sky that
includes the constellation Capricorn. In March/April of that
year, Capricorn rose above the eastern horizon when viewed
from Arabia and thereabouts.

The Magi had the knowledge and cultural influences that
would motivate them to chase the comet, says Humphreys. In
classical literature, the Magi are depicted as a religious group
skilled in the observation of the heavens. From the fourth
century BC, Babylon was the centre of astronomy in the known
world and the Magi were important members of the
Babylonian royal court in Mesopotamia. Moreover, Babylon
had contained a thriving Jewish colony since the time of the
Exile in the 6th century BC, so that the Jewish prophecies of a
saviour king, the Messiah, may have been well known to the
Magi.

In the Hellenistic age (322–30 BC) some of the Magi left
Babylon for neighbouring countries, and by the time of the
birth of Christ, they lived mainly in Persia, Mesopotamia and
Arabia (now Iran, Iraq and Saudi Arabia). Humphreys suggests
the Magi who saw the star of Bethlehem probably came from
Arabia or Mesopotamia.

Why did they follow the star? Comets were associated with
great rulers, and the Magi were known to have visited kings in
other countries. Not everyone agrees. Critics point out that
Ptolemy, the second-century astronomer/astrologer from
Alexandria, associated comets with misfortune. His *Tetrabiblos*
(the 'bible' of astrologers) warned: 'Through the parts of the
zodiac in which their heads appear and through the directions

in which the shapes of their tails point, are the regions upon which the misfortunes impend.'

Colin Humphreys points out that the comet was visible for seventy days, which is consistent with what is known of the Magi's journey: the distance from Babylon to Jerusalem was at most 900 miles, which would have taken between ten and twenty days (a fully-loaded camel can handle between 50 and 100 miles a day). In Jerusalem, the Magi would have discussed the significance of the heavenly portents with Herod, says Humphreys: 'The story started in May 7 BC when Jupiter and Saturn came together against a backdrop of Pisces, signifying that a Son of God would be born in Israel. They would have told him that this happened twice more in 7 BC to reinforce the message, and then they would have told of other events, a triple massing of planets in 6 BC and the comet in 5 BC, which conveyed the message that the birth was about to happen. Hence they jumped on their camels and came to Jerusalem.'

How did the comet direct the Magi to Bethlehem? Given the model of the heavens that then prevailed, comets would have been regarded by the Magi as being below the 'heavenly spheres' containing the stars, planets and so on. Humphreys explains how the Magi might have thought of the comet as hanging over a given spot, particularly if it was low in the sky and its tail was oriented vertically. This interpretation vividly fits Matthew's account: 'Lo, the star, which they had seen in the east, went before them, till it came and stood over where the young child was.'

There has, however, been some debate over whether the Chinese records imply movement typical of a comet. According to Humphreys, the phrase 'a *sui-hsing* appeared at' implies definite motion. Others have taken the translation literally and suggest that the Chinese saw a point of light suddenly 'come on' in the sky. The latter theory has led some British astronomers to suggest that the Chinese mistakenly categorized the object as a broom star when it was in fact a 'guest star', the thermonuclear flash of a nova, from the Latin *nova stella*, or 'new star'. This theory dates back a long way, perhaps even to a hint in *De Vero Anno*, written in 1614 by the great astronomer Johann Kepler. A few such novas appear each year, when a faint, usually unseen, star brightens by a factor of 10,000 or even 1,000,000. These outbursts are thought to occur in a binary, a pair of stars, when gases from the larger member fall into the smaller member, triggering a nuclear conflagration.

However, the same reasons that make the comet an attractive candidate for the Bethlehem star tend to disqualify the nova. Matthew 2:9 suggests that the object was later visible in the south, and a nova would not have moved that much. The location is also an unlikely one for a nova, given that the Bethlehem star appeared well away from the disc-like plane of our galaxy, which is lush with stars – its hazy cross-section is seen in the sky as the Milky Way – and likelier to be a stellar nursery.

A star death, or supernova, would provide another dramatic candidate, one with the potential to light up the night sky. Indeed, in AD 1054 Chinese astronomers observed a supernova that was bright enough to see in the daytime. However, this suggestion has also been discounted, since the remnant of such a cataclysm, close enough to the Earth for it to have appeared bright, would have left a spectacular aftermath, a splash of radio and X-ray wavelengths that would still be visible to astronomers today.

Rival star theorists dismiss both novas and comets, pointing out that Middle Eastern astrologers of the period were preoccupied with the planets, the Sun and Moon, and had little time for other heavenly bodies. If we accept only this narrow range of star candidates, other possibilities emerge. Perhaps the Wise Men were struck by a moment of hesitation on the part of Jupiter. In an article published in 1992 in the *Quarterly Journal of the Royal Astronomical Society*, the late British polymath Ivor Bulmer-Thomas proposed that the Bethlehem Star was Jupiter passing through a stationary point in its trek across the sky. When a planet undergoes such a 'retrograde motion', a consequence of the relative position of the planets around the Sun, it makes a loop against the stars. The Magi were seeking a king of the Jews because this motion had been executed by Jupiter, the most regal of all the planets.

Bulmer-Thomas goes on to argue that other celestial events, involving other star candidates, alerted the Wise Men to the stationary point. Three conjunctions of Jupiter and Saturn in 7 BC were considered highly significant by contemporary astrologers. The star almanac of Sippar, a clay tablet found about 30 miles north of Babylon, refers in detail to this triple conjunction, which was followed by the grouping in Pisces of Mars, Jupiter and Saturn in 6 BC, and by the comet of March/April 5 BC. Forewarned by these celestial events, the Magi would have followed Jupiter from the time it emerged from behind the Sun in May 5 BC. Bulmer-Thomas says that they would have seen Jupiter pass through a stationary point four months later, about the time it took for them to complete their journey: 'As they approached Bethlehem in the fourth week of September they could see that Jupiter was near its first stationary point, and this convinced them that the babe they saw lying in a manger was indeed the Messiah.'

THE MAGI AS ASTROLOGERS

The above discussion of conjunctions and retrograde motions makes the objective perspective of a modern astronomer an inadequate one from which to hunt for the Bethlehem star. Rather, we need to understand who the Wise Men were and how they interpreted signs in the heavens. Astrology was widely practised throughout the Roman world, especially in that part of the Near East that included Judea, and the Magi, with their detailed knowledge of the night skies, would have been unlikely to have been impressed by a routine event such as the appearance of a shooting star. They might, however, have been moved by something in the night skies that would seem unremarkable to a modern astronomer. This is best understood by looking back at the common origin of astronomy and astrology.

Before the seventeenth century, there was not the sharp dichotomy that we see today between astrologers (who always spout ambiguous rubbish) and astronomers (who sometimes do). At the root of both disciplines is our ancient fascination with the night sky. A holy man's knowledge of the heavens conferred an – albeit limited – ability to foretell the future, guiding him through the seasons, showing when to harvest and when to move herds. It also helped him to predict notable events such as a solar eclipse or the flooding of rivers like the Nile. In this restricted sense, knowledge of the heavens illuminates our future. This, however, is a far cry from the astrologer's supposed art of judging the occult influence of the stars on human affairs.

Woe betide anyone who confuses astronomy and astrology today. But when the Wise Men gazed at the heavens, they could be forgiven for thinking that they glimpsed something of their destiny. Once we accept that the Magi had an astronomer's interest in the details of the night sky, spiced with the astrologer's fascination for what these details might say about human affairs, then it becomes apparent they may not have seen a star at all, or indeed a cut-and-dried astronomical object, but an unremarkable cosmic event with remarkable symbolism.

This fascination with cosmic symbols underlines one clear difference between the Magi and the chief priests: astrology was practised in Babylonian society, while it was forbidden in Jewish society, according to Deuteronomy 4:19 ('lest ye corrupt yourselves … lest thou lift up thine eyes unto heaven, and, when thou seest the sun, and the moon, and the stars, even all the host of heaven, shouldest be driven to worship them, and serve them'). That Herod was unaware of the star until the

Magi informed him of its significance adds weight to this argument. (Those who prefer a straightforward astronomical interpretation would, of course, disagree and point out that there is nothing in the Bible to say Herod did miss the star).

CHRISTMAS ASTROLOGY

If we accept that many Bethlehem-star suggestions do not take into account the mind-set of the Wise Men, what kind of astrology was practised in the Near East during the reign of King Herod? Michael Molnar from Rutgers University in Piscataway, New Jersey, has studied Greek astrology as used throughout the Roman world, including Mesopotamia and Babylonia, and drawn his own conclusions: 'By my theory, Jesus would have been 2,000 years old on April 17, 1995.' His candidate for the star is an event that took place on 17 April 6 BC: a double occultation of Jupiter by the Moon, when our closest neighbour moves in front of the giant planet. Molnar's studies have suggested that this event, though of little significance to a modern astronomer, was 'brilliant' in an astrological sense.

Michael Molnar notes that astrological signs appeared on ancient coinage, notably from Antioch, the capital of the Roman province of Syria. On one side of each coin was a bust of Jupiter. On the other, Aries the ram gazed back at a star. Molnar now believes that the coins commemorate the annexation of Judea by the Romans, which suggests that the Romans were aware of important astrological portents involving Judea. He considers it likely that what he calls 'the great portent' of 17 April 6 BC was very much on their minds – the Romans were looking for proof that a Roman, not a Jew, had fulfilled the messianic prophecy. (They would indeed have thought this to be the case when, a dozen years later, Augustus Caesar assumed control of Judea). Aries appeared on the coins because it was linked to Judea in contemporary symbolism: Ptolemy mentions that Judea is under the spell of Aries. On this point, Molnar's theory conflicts with others that argue that Pisces is the sign of the Jews. However, Molnar counters that such theories rest on a Renaissance source, as opposed to Ptolemy's first-century BC *Tetrabiblos*. He adds that there are other sources from Roman times that also support Aries as the symbol of Judea.

Molnar's argument needed another ingredient – the presence of a heavenly body to symbolize the birth of a king: 'My initial search for a regal "star" centered on the star of Zeus, namely the planet Jupiter, which invariably played the central role in horoscopes that had regal implications.' He found that the regal symbol of Jupiter did indeed feature prominently in the ancient horoscopes of several Roman emperors. To identify an astrological portent involving Jupiter, he focused on lunar occultations. These are 'bull's-eye' conjunctions in which the Moon's disc obscures the planet. Examining the likely time frame, Molnar found only two that took place in Aries and thus in Judea, occurring on 20 March 6 BC and 17 April 6 BC: 'This finding was regarded as an intriguing coincidence until I later realized that during the second occultation, Jupiter was precisely "in the east," an astrological terminology that Matthew uses to describe the Magi's star.' The heavens on 17 April 6 BC produced impressive astrological portents: 'If we recreate an horoscopic chart for [this date],' writes Molnar, 'we find unmistakable indications pointing to the birth of a king of Judea. I believe that a horoscope of that day was incredibly ominous – truly messianic.'

The mystery of the star has been solved. Perhaps not. David Hughes of Sheffield University, for one, believes that such occultations took place too regularly to be of great astrological significance. ('How often do you want the Magi to go to Jerusalem?' he asks.) Hughes is most struck by the rival idea of a triple conjunction. One between Mars, Jupiter and Saturn was historically believed to have preceded the birth of Christ. In 1465 Jakob von Speyer, the court astronomer to Prince

Frederic d'Urbino, asked the German astronomer Johann Müller (Regiomontanus) the following question: 'Given that the appearance of Christ is regarded as a consequence of the Grand Conjunction of the three superior planets, find the year of his birth.' Müller was unable to answer. In 1604 the German astronomer Johann Kepler calculated that the massing of Jupiter, Saturn and Mars occurred every 805 years.

David Hughes argues that the Bethlehem triple conjunction was not of three planets but of Jupiter, Saturn and the constellation of Pisces. The regal aspect came from Jupiter, while Saturn stood for both the principle of justice and the land of Palestine. Pisces was the sign of the zodiac that represented the land of Israel. This conjunction, claims Hughes, signified a potent brew of divinity, kingship and righteousness involving the Jewish people and the Promised Land: 'Putting it crudely, that is why the Wise Men went for Jerusalem.'

The Magi could have figured out the details of the triple conjunction well in advance. They could have watched the first conjunction from Babylon in May of 7 BC, but delayed travelling until the end of the long, hot summer. On their way to Jerusalem, they could have witnessed the astrologically important moment when Jupiter and Saturn were rising at the instant of sunset. As interpreted by David Hughes, the passage rendered in most translations of the Bible as 'We have seen his star in the east' has a more specific meaning, namely, 'We have seen his star rising in the east as the Sun was setting.' If this explanation is correct, the only thing miraculous about Hughes's theory is that the Magi noticed the 'star' and made the arduous trek to witness, as they said, the appearance of a new king for the Jews. This suggests the real Christmas should be celebrated some time around September, to reflect the events that took place in 7 BC. However, given the patchy evidence, the Bethlehem star debate will no doubt continue.

WHY DECEMBER?

How, then, did the birth of Christ become entangled with winter and its paraphernalia of reindeers, Santas and so on? The origin of the December date lies in the festivals of pagan times, when people feared that winter might snuff out the sun god's rule, allowing the powers of darkness to take over. Illuminated by today's scientific knowledge, this primitive dread is under-standable: with the exception of strange life forms that huddle around volcanic cracks in the ocean floor, or bugs that dine on wet basalt, the living economy of the planet relies on harvesting light from the Sun.

The traditional time for celebration, 25 December, coincides with pagan festivals and follows closely on from the winter sol-stice, when the Sun reaches its greatest excursion south of the Equator. Late December marked an important turning point in the Sun's apparent course, as the daily quota of sunlight grew longer and stronger. Since ancient times, people have lit candles, bonfires and yule logs to help nourish the sun god when he was at his weakest and drive winter and its hardships away. The Roman festival of Saturnalia in mid-December is one example. In keeping with its name – *saturnus* means 'plenty' or 'bounty' – the celebration involved feasting, gam-bling, dancing and singing in honour of Saturn, the Roman god of agriculture. Hats were worn, though not paper ones. At these festivals, the master served the slave, a ritual that can still be seen at office parties today. Gifts were also given, including branches of sacred wood (these evolved into the switches left to punish bad boys and girls in some seasonal festivities). This fes-tival was swiftly followed by the Kalends, a new-year celebra-tion. Already we can see the genesis of the celebrations we enjoy today.

The pagan Roman emperor Aurelian proclaimed 25 December as *Natalis Solis Invicti*, the festival of the birth of the invincible sun, which was marked by chariot racing and decorations of branches and small trees. In pagan customs pre-dating even Roman times, evergreens such as holly and ivy were used for decoration and presents. These plants stay fresh-looking during the winter months so were thought to be linked to wood spirits and were associated with vitality. As with other primitive rites, these customs were assimilated into subsequent traditions. The thorns of the holly, for example, were equated with Christ's crown of thorns. Kissing under the mistletoe is another custom that dates back to pagan times, when the plant was associated with fertility. When it was found growing on oak, the sacred tree of the Druids, it was treated with special reverence.

The ancient Britons and Scandinavians also held mid-winter sun festivals, which they called 'yule' or 'jol' (this is where the words 'yuletide' and 'jolly' come from). Nor did the Germanic tribes of northern Europe miss out on the fun. They assembled in midwinter for feasting, drinking, religious observances and simple merriment. White-bearded Odin (sound familiar?) was thought to roam around punishing evildoers during the cele-bration in December, prompting attempts to appease him with gifts.

These pagan midwinter festivals remained popular centuries after Christ was born and early Christians were unwilling to relinquish them. When the Church found it impossible, despite repeated bans, to abolish all pagan customs, it Christianized a number of them, divesting them of their worst features. This created another problem. Since no one knew the date of Christ's birthday, some people celebrated it in the spring and others in the winter.

The first mention of Christmas Day, as far as we know, was in the Roman calendar Chronographus Anni CCCLIV (chron-ographer of the year 354). Within half a century, Christmas Day had become an important date in the Christian year, with 25 December fixed as the 'natural' date so as to exorcize the earlier pagan festival of the winter solstice. The short form for

Christmas, Xmas, started out as an ecclesiastical abbreviation that was used in tables and charts; the first letter of the word Christ in Greek (*khristos*) is *chi*, which is identical to our X. Thus X stands for Christ, as in Xmas.

Those who believe in star signs would, no doubt, be fascinated to know the precise date of the birth of Christ and whether he behaved like a typical Capricorn, Aries or Pisces. Recent research has provided some fascinating insights into the way in which the timing of Jesus's birthday might have influenced the rest of his life, from the functioning of his immune system and chances of getting heart disease to his potential for intellectual and sporting achievement.

The last-mentioned study, concerning physical prowess, was carried out at the University of Amsterdam. The results suggest that if soccer had been played in Judea two millennia ago, Jesus would have had a better chance of making his local team if he had been born between August and October. This weak 'astrological' effect is most notable in sports where physical development matters: the youngest children in any age grouping are at a disadvantage because they are born late in a selection year.

Unfortunately for Jesus, another study, this time into the connection between birth dates and academic success later in life, showed that it helps to be born during the summer. Takanari Gotoda of the Hammersmith Hospital, west London, came to this conclusion after trawling through the birth dates of 2,525 graduates from the faculty of medicine at the University of Tokyo, many of whom achieved outstanding results in their high-school examinations. He grouped them by the month of their birth and then divided this number by the number he would have expected given the monthly birth records in Japan over the same period, from 1947 to 1971. What emerged was a hump-shaped plot with its peak in the summer, suggesting that the summer-born are the highest academic achievers.

Although the reason for this seasonal effect on intelligence are not clear, you do not have to look to astrology for a possible explanation: babies born during the warm summer months are less likely to be wrapped up, suffer illness or be cooped up indoors. This would allow them to be more adventurous and enjoy a more stimulating environment, which has been shown by experiments on animals to be important for brain development.

Research by Michael Holmes of Queen Margaret College in Edinburgh suggests that revolutionaries tend to be born around Christmas (they have birthdays between October and April), whereas May to September is the season for reactionaries. In the days before central heating, Holmes argues, the winter-born proto-revolutionary would have found freedom to explore with the coming of summer. 'The summer-born non-revolutionary, on the other hand, would have enjoyed more freedom at first, when less able to use it, but would have been constrained the following winter when ready to extend his explorations.' Although Takanari Gotoda's work suggests that the summer-born may be smarter, it appears they are also less adventurous. Whatever the outcome, Holmes's discovery seems somehow more appropriate than Gotoda's, given the impact of the birth of Christ and his official December birth date.

CHRISTMAS CARDS GET IT WRONG (AGAIN)

The Christmas star is only one aspect of the dubious sentimentalized account of the Nativity, in which the Holy Family arrive in Bethlehem late at night and there is no room at the local inn. The exhausted couple are forced to stay in a stable, where Jesus is born. This popular version of events is refuted by Luke's Gospel, argues Ken Bailey, Director of the Institute for Middle Eastern New Testament Studies in Beirut. The Gospels translate the Aramaic speech of Jesus and his disciples into Greek. The careless translation of key words of biblical Greek has led to the misinterpretation of the story from a modern, Western perspective.

The Bible tells us that Mary gave birth to her first son, wrapped him in swaddling clothes and laid him in a manger. The traditional interpretation of this story, in the West at least, is as follows: Jesus was laid in a manger. Mangers are naturally found in animal stables. Ergo, Jesus was born in a stable. But this is to overlook the traditional design of peasant homes in Palestine and Lebanon, says Ken Bailey. Farm animals would share the same living area as the family so the beasts could help warm the house in winter and at the same time be safe from thieves. Sometimes there was also an adjoining guest room. The family would live on a raised terrace (called a *mastaba* in Arabic) in one room, while their 'central heating' – oxen, donkeys and so on – would have their place some 4 feet below on the actual floor (*ka'al-bayt*) near the door. The mangers, often hollowed stone filled with crushed straw, would be found on this floor, or at the edge of the terrace. This reinterpretation fits in with the idea of Jesus being born in poverty, says Bailey: 'That is, he is born in a simple peasant home with the mangers in the family room. He is one of them.'

Bailey also questions whether the local inn was really full. He argues that the word 'inn' is a mistranslation of *kataluma*, which has several meanings. It can mean 'inn' but it can also mean 'house' and 'guest room'. Reading *kataluma* as 'inn' creates several problems. First, Luke uses another word – *pandokheion* – for a commercial inn. Second, the only other use of the noun *kataluma* is in Luke 22:11 and Mark 14:14, where 'guest room' better suits the context ('And ye shall say unto the good man of the house ... Where is the guest chamber, where I shall eat the passover with my disciples?'). Third, another instance of cultural misinterpretation, if Joseph had stayed at an inn he would have insulted any members of his extended clan who still lived in his family's native village and were duty bound to offer hospitality to kin. Finally, there is great uncertainty over whether Bethlehem had a commercial inn at all.

Bailey argues that *kataluma* should be translated as 'guest room'. He draws on cultural and archaeological evidence to support his contention that Jesus was born in the heart of a Palestinian home: 'Joseph and Mary arrive in Bethlehem; Joseph finds shelter with a family; the family has a separate guest room but it is full. The couple is accommodated among the family in acceptable village style. The birth takes place there on the raised terrace of the family home and the baby is laid out in a manger.'

As in the case of the Wise Men, we may have made the mistake of taking literally a Nativity story that has been interpreted according to Western culture and language. Through the eyes of a Palestinian, the verse makes sense on a cultural, historical and linguistic level. St Luke writes: 'And she gave birth to her first-born son and wrapped him in swaddling clothes, and laid him in a manger.' A Palestinian would instinctively think, 'Manger – oh, they're in the main family room. Why not the guest room?' The writer anticipates the question and replies: 'because there was no place for them in the guest room.' The reader would conclude, 'Ah, yes – well, the family room is more appropriate anyway.' Denis Alexander, Editor of the journal *Science & Christian Belief*, finds this reinterpretation convincing: 'If [Ken Bailey's] interpretation is correct (and he certainly persuaded me) then the pictures on many millions of Christmas cards (Mary and Joseph being shut out of the Inn and so on) are simply mistaken, not to speak of a few million sermons.'

THE FIRST CHRISTMAS PRESENTS

We know that the Wise Men came with presents of gold, frankincense and myrrh to mark the birth of the new king (given that they were astrologers, they probably cast his horoscope as well). However, the modern reader may be puzzled by the gifts of the Magi – gold is obvious, but why frankincense and myrrh? These resins, collected from trees for more than

5,000 years, make sweet-smelling perfume and incense but were also reputed to ward off all manner of ills. They were in short supply due to intense demand, which is why they ranked with gold as gifts suitable for the Christ child.

Frankincense (from the French, meaning pure incense) has long been valued for the sweet-smelling fumes it produces when burnt. It comes from a spiky tree, *Boswellia carteri*, found in the dry areas of north-eastern Africa and southern Arabia, such as the arid highlands of Somalia and the Arabian peninsula. The resin is harvested by nomadic tribes, who scrape the bark of the tree and return months later to collect the 'tears' of solidified whitish resin.

Myrrh is a yellowish-red resin collected from the short, thorny tree *Commiphora myrrha*, which grows across Ethiopia and Kenya and produces oily, bitter-tasting resin at the bases of its branches. Other species of commiphora also produce resins, for example *C. abyssinica* (African myrrh) and *C. mukul* (guggulu).

These fragrant resins are all remarkably similar in biochemical terms and are probably produced by the trees to gum up the mouthparts of attacking termites and to aid repair through their antibiotic properties and by acting as a temporary dressing for damaged bark. The ancient Egyptians used frankincense to treat wounds and also in religious rites, when anointing mummified bodies. During Roman times, when cremation was widely practised, it was customary to burn frankincense in the funeral pyre; according to Pliny the Elder, this was done not just to appease the gods but also to disguise the grim odour. Between AD 25 and 35, Celsus, a Roman author, recommended frankincense for treating cuts and bleeding. In his medical encyclopedia he suggested it as a possible antidote to poisoning by hemlock. In the seventeenth century, Nicholas Culpeper, a herbalist and apothecary in Spitalfields, London, used frankincense to treat stomach ulcers and bruises.

What about myrrh? Early Egyptian myths describe the resin

as the 'tears of Horus', the god of the sun and moon. Ancient Egyptians used myrrh as well as frankincense in mummification – its antibiotic qualities reduced decay and helped to prevent tissues falling apart, and its aroma disguised the smell of the body. It was also used for this purpose by the living. Queen Hatshepsut of the 18th dynasty rubbed it into her legs. Early Sumerian inscriptions describe myrrh treatments for bad teeth and worms, while the Greeks prescribed myrrh or its oils for infections of the mouth, teeth and eyes, as well as for coughs. Both Greeks and Romans thought that myrrh could cure poisoning by snake bites. The use of a little myrrh in 'aromatic' wines was found to prolong shelf life.

Modern studies have discovered all these resins to have antiseptic, antifungal and anti-inflammatory properties, and therefore can make valuable dressings. The oils made from the resins seem to cause the bronchi of the lungs to dilate, and may help to relieve lung infections and mild asthma. More striking still, researchers have found that eating resin or oil from *C. mukul* or guggulu lowers blood-cholesterol levels, while Chinese studies on animals have shown that *C. myrrha* can reduce the development of arteriosclerosis, or hardening of the arteries.

Recently, scientists at the University of Florence, Italy, led by Piero Dolara, have undertaken research into myrrh, based on the confused references to its properties and uses in ancient and biblical texts. Laboratory mice were given either a control or myrrh and were placed on an uncomfortably warm metal plate. Comparison of the paw-licking times of the two groups showed that myrrh is a painkiller, with an action similar to that of morphine. Follow-up studies identified two types of sesquiterpenes, chemicals having analgesic effects. This experiment on mice perhaps explains the reason why *vinum murratum* – wine with myrrh – was offered to Christ shortly before the Crucifixion. Modern science suggests that Jesus was being offered a primitive anaesthetic.

two

Miracle

He came all so still
Where his mother lay,
As dew in April
That falleth on the spray.

Mother and maiden
Was never none but she;
Well may such a lady
Godes mother be.
    Anon. (fourteenth century)

Anyone with even the haziest grasp of biology would be troubled by the traditional account of the Nativity. Even in the earliest times some heretical sects, such as the Philanthropists and Adoptionists (who had unusual ideas about the divinity of Jesus), questioned the idea that Christ had no father and that he was conceived in Mary by the power of the Holy Spirit. Today, although millions of people send cards adorned with the classic image of the Virgin Mary, many are sceptical that she remained a virgin despite conceiving Jesus.

This unease has been expressed at a very senior level in the Christian Church. The former Bishop of Durham, David Jenkins, caused a furore by pronouncing the Virgin Birth a myth. One Moderator of the General Assembly of the Church of Scotland delivered the strangulated statement: 'It doesn't seem to me impossible that God could not choose a fully-human birth as a way of entering the world.' His pronouncement, though equivocal, resulted in an open letter signed by more than 100 Church of Scotland ministers, which declared: 'We believe unreservedly in the historicity of the Virgin Birth and regard it as integral to our faith in the incarnate Christ.'

The debate centres on more than mere scientific possibility. On a theological level, some argue that a virgin birth seems to separate Jesus from the rest of humankind. Others reason that it is theologically necessary since Jesus is divine as well as human. Indeed, some would argue that if the Christ child was to be truly man and truly God, it would be surprising if he did not mark his continuity with humankind by normal birth after normal gestation, while at the same time emphasizing his discontinuity through a device such as a virgin birth.

Let's examine the biblical evidence for the birth. Matthew (1:18, 22–3) and Isaiah (7:14) tell us that 'Mary was betrothed to Joseph; before their marriage she found she was going to have a child through the Holy Spirit ... this happened in order to fulfil what the Lord declared through the prophet: "A virgin shall conceive and bear a son, and he shall be called Emmanuel," a name that means "God with us."' Mary's reaction was doubtful, according to Luke (1:26–35): 'How can this be?' she complained to the angel Gabriel. 'I am still a virgin.' The angel replied: 'The Holy Spirit will come upon you, and the power of the Most High will overshadow you; for that reason the holy child to be born will be called the Son of God.'

Scientifically, it would have been so much easier to accept the Virgin Birth if Jesus had been born today. A huge gamut of reproductive science and technology could be called on, in particular the newly established technique of cloning. The idea that an adult human could be cloned from a single cell became a real possibility in 1997 with the announcement of the world's first clone of an adult animal: a Finn Dorset sheep called Dolly, developed by the Roslin Institute of Edinburgh and neighbouring PPL Therapeutics. Ian Wilmut and his colleagues found a way to take cells from the tissue of an adult (in this case, from the mammary gland of a sheep), mass-produce them and then extract the part containing the genetic code (the nucleus) and transfer it into an egg that had had its nucleus removed. The resulting 'reconstructed embryo' was given a jolt of electricity to trigger cell division, then implanted into a surrogate mother. The last steps can be repeated again and again to make endless copies.

THE VIRGIN BIRTH

Such scientific possibilities are not necessarily incompatible with religious ideas. Even if the mechanics of the Virgin Birth were satisfactorily explained, this would not preclude a divine plan which, for modern Christians, is the true miracle of Christ's birth. The scientist Sam Berry has added a new twist to the debate. A committed Christian who believes in miracles, and a professor of genetics at University College London, he has proposed several unlikely, though biologically plausible, mechanisms by which the virgin birth of a male child could occur. He is not suggesting that God necessarily used any of these virgin-birth scenarios, but is simply pointing out that one should not discount the traditional view on scientific grounds: 'The revisionist scepticism typified by David Jenkins is well known and, although driven by a legitimate search for the reality of God, has the danger of throwing the baby out with the bath water.'

Berry argues that it is right to examine faith and dogma using all our scientific insights. He declares that 'Baseless credulity is a sin – a disservice to the God of truth.' He echoes the comments of the physicist and Anglican priest John Polkinghorne, who said: 'The test of credibility will lie in whether one can articulate a coherent understanding of the world in which such phenomena can find a fitting place.'

In formulating his ideas, Sam Berry can call on the expert advice of his wife Caroline, a genetics consultant at Guy's

Hospital in London. He has also been a member of the Human Fertilization and Embryology Authority, which licenses and regulates British clinics that carry out donor insemination and *in-vitro* fertilization. Furthermore, he has outlined his ideas in the journal *Science & Christian Belief*, whose editor is Denis Alexander of The Babraham Institute in Cambridge. 'It is the first time I have come across anything by a serious scientist suggesting that it is not scientifically implausible,' says Alexander. 'He has put together a number of recent findings to make an interesting story.' Berry's paper was, he said, referred by two experts in the field, 'neither of whom believe in the Virgin Birth.'

However, this is not to say that Sam Berry's ideas have been accepted by his peers. A critique by the theologian Peter Addinall, published in the same journal, contains a different interpretation of the biblical references to the birth of Christ. This scholarly argument is reminiscent of the one that has dogged attempts to identify the Bethlehem star. Addinall points out that Isaiah 7:14 refers to a 'young woman', and not all young women are necessarily virgins. But Berry replies that the context makes it quite clear that 'virgin' is more appropriate than 'young woman'; it would hardly be worth recording that a 'young woman' had become pregnant.

As was the case with the manger in the Nativity story (see Chapter 1), cultural misunderstanding can lead to confusion. Middle Eastern languages today use words like *kiz* (Turkish) or *bint* (Arabic) for 'young woman'. There is a strong implication that the well-chaperoned *kiz* or *bint* is a virgin but this probably reflects the expectations of traditional Middle Eastern cultures. Feminists would add that, when it comes to the way language reflects the interests of a patriarchal society, English is not far behind: 'maiden' is synonymous with both 'girl' and 'virgin', and its disappearance from everyday use reflects changes in the perception of women.

Addinall also claims that Berry's attempts to explain the

Virgin Birth create a 'fatal ambiguity' in our definition of what is miraculous. He argues that science is absolutely irrelevant to miracles, which defy human understanding: 'Miracles which can be explained in naturalistic terms, even when we are dealing with an extremely remote possibility, cease to be miracles, except in the weak subjective sense of being events which cause amazement. We are left with an historical incident which modern scepticism is compelled to accept, but with no evidence that it was actually an instance of direct divine intervention in the world's working.'

Berry's reply is a refusal to accept a distinction between miracles that can be 'explained naturalistically' and those in which 'God alters the course which nature, left to itself, would follow.' Just because something is a miracle, that does not mean that you cannot explain it, he argues. If God is continuously sustaining all the processes of creation, as Christians believe, then a 'miracle' is simply an unusual activity of God. As Denis Alexander puts it, 'The Bible sees the universe as a seamless cloth of God's continuing creative activity (which scientific laws attempt to describe, albeit inadequately), so miracles are the wrinkles on that cloth.'

On occasion, the Bible provides explicit mechanistic explanations for miracles in order to reveal God's concern for people. Berry cites as an example Exodus 14:21–22, where 'The Lord drove the sea back by a strong east wind all night, and made the sea dry land, and the waters were divided. And the people of Israel went into the midst of the sea on dry ground.' The actual site of the Israelite crossing is uncertain but one suggestion is that it was not the Red Sea but a tract of low-lying land on the Nile Delta called the Sea of Reeds or the Reed Sea. (This distinction is made in a document preserved among the Dead Sea Scrolls, known as the Genesis Apocryphon, which suggests that the Reed Sea is an inlet of the Red Sea.) This area was passable most of the time but was known to be flooded to a considerable depth by freak tides. It is feasible that the Israelites got across with no more than wet feet, as the wind delayed the normal flow of water, while the pursuing Egyptians were swept up by the flood tide. 'The miracle lay in the place and timing of a physical event, not merely in the fact of its occurrence,' says Sam Berry.

A number of other biblical events are equally amenable to what a scientist would call a reductionist, materialist explanation. Obeying God's command, Moses turns the River Nile blood red (Exodus 7:19–21). Red-clay deposits from upstream lakes in Ethiopia not infrequently stain the Nile flood waters a reddish-brown colour similar to blood, and also stir up plankton called dinoflagellates (named after their whip-like 'tail'), which produce red tides that are toxic to fish.

The plagues of Egypt are also explicable. Prolonged flooding can lead, and has done so, to population explosions of frogs and biting insects; flies often transmit epidemic diseases of domestic animals ('all the cattle of the Egyptians died'). Similarly, locusts and sand-storms ('darkness') are common in the Near East.

EVOLUTION AND VIRGIN BIRTHS

Miracles aside, the Virgin Birth raises another issue dear to the hearts of scientists: why did sexual procreation evolve at all? Our most primitive ancestors emerged some 3,850 million years ago, but sex itself began only about 1,000 million years ago. Before then, all creatures presumably existed as clones.

In his book *The Evolution of Sex*, John Maynard Smith of Sussex University maintains that there is a gap in our basic understanding of why sex evolved. 'I have been wondering about this for fifty years but I don't claim to have solved the problem,' he told me. 'The problem has been that, in the short run, abandoning sex would be an enormous advantage, at least for females.' Calculation illuminates the joys of reproducing without sex: sexually reproducing females on average produce one female offspring, while asexually reproducing ones, who rely on virgin births, will produce two. Thus, among a colony of asexually and sexually reproducing individuals, the former would quickly dominate. So why is it that females go through all the fuss and bother of finding a mate and then dilute their genes with those of a male in any resulting offspring? Sex must offer advantages, otherwise we would not be here to consider the question.

Scientists have come up with various answers. First of all, sex introduces variety. It shuffles genes more effectively than asexual reproduction, introducing more genetic variability into a population and enabling it to cope with change, such as the arrival of a new disease or parasite. For example, a rust fungus introduced in Chile ravaged an asexually reproducing black-berry while leaving the sexually reproducing variety almost unscathed. Another suggestion, called the 'engine and gearbox' idea, is that sex helps us to cope with bad mutations. It goes like this: if you buy two clapped-out motor cars, one without an engine and the other without a gear box, and combine them, just as sex mixes male and female chromosomes, you can produce one functional vehicle. Mathematical models confirm this, though the reason sex helps us cope with bad mutations is more subtle than the above example suggests. Remember, after all the story of the dancer Isadora Duncan, who wrote to George Bernard Shaw suggesting they should get together because, with her beauty and his brains, their children would be unrivalled. But what, Shaw replied, if the children had her brains and his beauty?

An alternative explanation of the emergence of sexual reproduction relies on the idea of the 'selfish' gene. Propounded by Richard Dawkins of Oxford University, this theory sees evolution in terms of genes that are concerned only with their own survival. A gene in an organism that relies on asexual reproduction would be confined to the descendants of that organism. But if it indulged in sex, it might get into other organisms' descendants as well.

To answer the puzzle of sex, scientists hope to get clues from organisms that forgo sex altogether, such as bdelloid rotifers, microscopic creatures that have a ring of whip-like feeding fronds. So far, no one has seen a male, let alone a hint of cryptic rotifer sex. Other evidence will come from studies of creatures that manage both methods of reproduction. The fact that greenfly and water fleas resort to sex now and again suggests there must be short-term benefits. 'It is a very sensible thing to do if you are an aphid,' says Maynard Smith. 'You can reproduce rapidly during the good times, without the bother of sex, and then have one bout of sex a year, which is quite enough to cope with the problems of rapid evolution and bad mutations.'

Slugs' strange sexual habits may also provide insights into why sex evolved. Being hermaphrodites, each one has both male and female sex organs, so a number of sexual permutations are possible. Research by Sam Berry's colleague Steve Jones and his team at University College London has discovered that the further north slugs are found in Britain, the more likely they are to forgo a sexual mate. It is the same with altitude: valley slugs reproduce sexually, unlike their self-fertilizing cousins in the mountains. These findings seem to offer support for the suggestion that sex enables a population to cope with change, according to gastropod-sex expert Les Noble of Aberdeen University. Sex is necessary when the creatures' environment alters in some capricious way, and this tends to occur in the warm south and in valleys, where there are more diseases, parasites and potential competitors. That, Noble explains, is when sexual reproduction is beneficial to a species: it throws up equally capricious and novel combinations of genes the parasites have never seen before and so cannot overcome. Meanwhile, atop Scottish mountains, life is more boring and predictable.

VIRGIN SCIENCE

Although this research is not yet complete, the experience of forlorn mountain slugs seems a long way from the Virgin Birth, which remains too much for many people to accept. Leaving aside the question of illegitimacy, raised in Matthew 1:19 ('Then Joseph her husband, being just a man, and not willing to make her a public example, was minded to put her away privily ...'), virgin humans simply do not have babies. Scientists, however, are not content to accept the lack of evidence of virgin births in humans, and seek underlying factors to explain why women are incapable of such births. The laws of nature do not forbid virgin births: eggs can develop of their own accord, for example, in bees or aphids, by a process called parthenogenesis. This is the case for around one in every 1,000 species. However, from a scientific viewpoint, if a virgin birth could have occurred, Jesus should have been a girl, not a boy.

The key limitation governing virgin births is that the genetic recipe for the offspring must, of course, come from the mother alone: in the case of Jesus, all his genes must have come from Mary. Under normal circumstances she would only have the genetic wherewithal – in humans, a bundle of genes called the X-chromosome – to make a female. For Mary to have given birth to a boy by parthenogenesis, she would also have had to have a Y-chromosome, a package of genes that separates the girls from the boys. Everyday sex turns the gender of a child into a lottery by mixing the chromosomes of each parent. This lottery takes place when egg greets sperm. An egg contains the mother's genetic instructions and an X-chromosome. Girls occur when the sperm adds an X-chromosome to the X already present in the egg, and boys when the sperm adds a Y-chromosome to the X-chromosome in the egg.

We can better understand why boys are conceived by study-ing the genetic cargo of the Y-chromosome. All chromosomes, X or Y, contain a double helix of the chemical DNA, tightly twisted into coils within coils. This chemical medium is the genetic 'message', spelling out genes by different sequences of four chemical units or 'letters'. For the purposes of the Virgin

Birth, the most relevant gene is one on the Y-chromosome called SRY. This triggers the construction of male sex organs and so on. If Mary had passed on a Y-chromosome, it suggests that she herself carried a working Y-chromosome. This creates difficulties for the scientist, since it would have led to her possessing male characteristics as well as being sterile. Nonetheless, the above discussion does admit the possibility that a baby girl could result from a virgin birth.

Of course, there are women who want to have a child without sexual intercourse. Biologically, however, this so called 'virgin-birth syndrome' is non-existent: insemination that gives rise to fertilized eggs and then offspring cannot be counted as a virgin birth. What we need is parthenogenesis. If King Edward potatoes, bees and greenfly can reproduce without males, could it also happen in human beings?

In 1955, Helen Spurway, a geneticist at University College London and wife of J.B.S. Haldane, the legendary Marxist biologist, undertook a study of human parthenogenesis, inspired by tropical fish. Spurway had bred a colony of guppies, a species of live-bearing aquarium fish, in which apparently virgin females were giving birth to young fish in the absence of males. This prompted her to speculate about human virgin births. If a virgin birth were to occur in our species, Spurway reasoned that such births must result in female children, through the lack of the Y-chromosome. Moreover, she believed that the claims of some women to have had a child by virgin birth could be vindicated by genetic tests. This prompted the *Sunday Pictorial*, a now-defunct British newspaper, to advertise for any mothers who genuinely believed that they had given birth to a parthenogenetic child.

Eleven of the nineteen claimants who came forward were quickly eliminated. They had been under the impression that a virgin birth meant simply that the hymen remained intact after conception. The other eight mother–daughter pairs were tested to see how genetically similar they were. If parthenogenesis had occurred, they should have been identical. One pair proved identical but for one test, involving skin grafts, when a degree of incompatibility was found. Tantalizingly, scientists of the day lacked the kind of tools used by modern molecular biologists that could have compared the genetic identity of mother and child in detail. 'Parthenogenesis may possibly have occurred in this case, although it has to remain unproven,' comments Sam Berry. As a footnote to this story, it emerged that the apparently virgin guppies that inspired Spurway were not normal females but contained some testicular tissue. This means that they were true hermaphrodites – a rare abnormality in guppies, as indeed in humans – and the offspring arose as a result of self-fertilization, a process that is quite different from parthenogenesis.

Sometimes an *un*fertilized human egg will begin dividing. This self-activated 'embryo' will create rudimentary bone and nerve, but seems unable to make some tissues, such as skeletal muscle, preventing further development. The end product is a strange form of tumour, a blend of hair and teeth known as a dermoid cyst, or teratoma. Because human parthenogenetic development never gets further than this, it suggests that barriers to development without a father were set early in mammalian evolution. The most profound of all has only recently been recognized: it is a phenomenon called imprinting.

In addition to the X- and Y-chromosomes that distinguish the sexes, there are twenty-two pairs of other chromosomes in humans, each carrying around 30,000 genes, each of which contributes a protein to the recipe of a human being. The process of imprinting ensures that the developing embryo relies on genes from both the mother and the father. In the early eighties scientists realized that genes from the mother's chromosomes do some jobs, while genes from the father's perform others. To understand what is going on, imagine your body as a bustling metropolis, sustained by many and various restaurants. At the heart of each is a cacophonous kitchen containing volumes of a vast, ancient cookbook. Every restaurant relies on the same cookbook, though numerous menus are created by chefs who specialize in particular cuisines. Each uses a different selection of recipes from the same cookbook – some prepare hot dogs and others green curry.

This kitchen metaphor describes what is going on in the body. The recipes are genes, the cookbook chromosomes and the chefs are transcription factors, proteins that bind to DNA with the effect of switching genes on or off. Together, they hold part of the answer to one of biology's great puzzles: how the clutch of identical cells in the early embryo, all with the same genetic blueprints, develop into distinct tissues, from skin to nerve to liver, each dominated by the actions of distinct subsets of genes. Among these subsets are imprinted genes, a small but important minority that is 'marked' in such a way that identical genes inherited from a person's father and mother function differently. Imprinted genes carry a biochemical label that reveals their parental origin and determines whether or not they are active inside the cells of the offspring. One is suppressed and inactive while the other is fully functional, depending on the parent who donated them.

Jesus would have found it tricky to make do with one parent's genes. Imprinted paternal genes have been found to be responsible for the development of the placenta, while imprinted maternal genes are involved in the growth of the embryo. Two Cambridge University researchers, Eric 'Barry' Keverne and Azim Surani, a pioneer in the field, have evidence, based on studies of mice, that the mother's genes contribute more to the development of the 'thinking' centres of the brain, while paternal genes have a greater impact on the development of the 'emotional' limbic brain.

Another example of the human dependency on genes from both the mother and father can be observed when imprinting conspires with a quirk of inheritance to produce an individual who lacks the active version of a gene. Prader-Willi syndrome, which causes mental handicap, obesity and other growth abnormalities, occurs in people who inherit both copies of chromosome 15 from their mother, rather than one from each parent. Paternal genes of chromosome 15 are required for normal development. Prader-Willi syndrome has a partner: Angelman's, causing mental handicap, characteristic jerky movements and epilepsy. Angelman's is also linked to chromosome 15, only this time sufferers lack functioning genes from their mother.

The importance of imprinting is further emphasized by the work of David Skuse of the Institute of Child Health, London, who found evidence of what he called a 'feminine-intuition' gene (or perhaps genes) that is imprinted on the X-chromosome. Working with Nina James of the Wessex Regional Genetics Laboratory, Salisbury, he discovered that the gene that fathers pass on to their daughters is active; the gene that mothers pass on to their sons is inactive. This ensures that normal girls are born with the active 'social-skills' gene, whereas normal boys are not. While men have to learn social skills, women are programmed to pick them up from birth, rather in the way that we are programmed to acquire language. What is more, the work of Skuse and James suggests that women inflict the curse of social inadequacy on men by 'switching off' the gene in their eggs before passing it on to their sons.

These examples show how imprinting would have made it tricky for Mary alone to have equipped baby Jesus with all the necessary genetic machinery. Complete failure of imprinting, and thus of any co-operation between maternal and paternal genetic instructions, seems likely to be lethal: as far as we know, no mammal conceived by parthenogenesis has ever been born in the wild (though some of these imprinting rules may have been broken by the cloning methods used to make Dolly the sheep).

However, there is a further possibility, hinted at by a real-life virgin birth, not of a baby, but of a cell line. This was found in a genetic halfway house called a chimera. When biologists talk about chimeras, they do not mean the fabulous monsters of Greek mythology (like the one with a lion's head, a goat's body and a dragon's tail, described by Homer). In biology, a chimera is a body made up of a mixture of various cell types. The chimera in this case had some cells with the usual combination of genetic material from mother and father, and others that appeared to be derived from the mother alone. It was investigated by the biologist David Bonthron and his colleagues at the University of Edinburgh using techniques from modern molecular biology that were not available when the *Sunday Pictorial* launched its quest for parthenogenesis four decades ago. Their report describes a one-year-old boy with a mild learning disability, slight facial asymmetry and small testes. His blood bears no trace of his father, although he must have had a Y-chromosome from his father in some cells to develop as a male. The boy, known in the scientific literature as 'FD', is thought to have been conceived when a sperm fertilized an egg that had already started dividing, rather than a single-celled egg as normally occurs. As a consequence, his blood system developed from egg cells that had not been fertilized, marking an extremely rare example of partial parthenogenesis. This, according to some observers, is the closest thing to a human virgin birth that modern science has so far recorded.

Experiments with mice have shown that parthenogenetic cells grow more slowly than normal cells and that the two can co-exist in the same tissue. The proportion of parthenogenetic cells in a given tissue type can also vary throughout the body. The researchers at the University of Edinburgh believe this could explain why FD's face is slightly asymmetric, with features smaller on the left-hand side. Bonthron notes that one in every few hundred people has slight asymmetry, and it is possible that, in rare cases, some could also be partially parthenogenetic.

However, we need more than parthenogenesis to account for the birth of Jesus. We also need a Y-chromosome to produce the Son of God rather than a daughter. In the absence of a sperm to import a Y-chromosome, Sam Berry speculates that Mary could actually have been male but have suffered a genetic mutation – called testicular feminization – that had the effect of preventing target cells in her body from 'recognizing' the male sex hormone testosterone. Mary would have been chromosomally male (XY) but would have appeared to be a completely normal female. In such a case, the female would be expected to be sterile and to lack a uterus. However, Berry points out that due to variations in the differentiation of the sex organs, it is possible that a person of this constitution could develop an ovum and a uterus: 'If this happened, and if the ovum developed parthenogenetically, and if a back-mutation to testosterone sensitivity took place, we would have the situation of an apparently normal woman giving birth without intercourse to a son.'

The possible scientific explanations for the Virgin Birth are by no means exhausted. For example, Sam Berry notes that some men are apparently XX, that is, have a female genetic complement: 'Examination shows that, in them, the male-determining factor of the Y [the SRY gene we encountered before] has been translocated onto another chromosome.' If this male-determining gene was translocated to the X-chromosome and if that chromosome was inactivated in early development (one X-chromosome is always inactivated on a random basis, so normally half the cells will express the maternal X, and half the paternal X), the carrier would have a female appearance, but have the capacity to pass on the male-determining gene. Men with XX-chromosomes are sterile. However, with enough ingenuity, this need not present a problem. 'Jesus never married and we do not know if he was fertile – although he was, of course, "perfect man" in the theological sense,' says Berry.

Does this settle the matter of the Virgin Birth? 'I must emphasize that there is no certain record of parthenogenesis in humans, nor of a male being conceived without fertilization by a Y-bearing sperm,' says Sam Berry. 'My point is that the possibility is not completely outside the realms of biological imagination. It would be wrong to extrapolate from the existence of an embryological pathway to an assertion that God worked in a particular way. That is not my intention. The mechanisms I have outlined are unlikely, unproven and involve the implication that either Jesus or Mary (or both) was developmentally abnormal. My purpose in describing them is simply to reduce the assumption of incredibility that seems to dog the doctrine of the Virgin Birth.'

VIRGIN ON THE RIDICULOUS

Unfortunately for Berry, this incredulity persists among other scientists confronted with his ideas. 'I am not sure that it helps with belief in the Virgin Birth, which after all is meant to be a miracle,' comments Wolf Reik of the Babraham Institute in Cambridge, who has studied the link between the failure of imprinting and cancer. 'I would think that biologically it would be a very complicated thing to achieve. If you string a lot of unlikely events together, you can explain almost anything.'

Scepticism was also expressed by Robin Lovell-Badge of the National Institute of Medical Research, Mill Hill, London: 'I don't believe that there is a rational scientific explanation. It would be going against pretty much everything we know.' Lovell-Badge has a special interest: he was one of the scientists who made the key breakthrough in the discovery of the SRY gene when, in 1990, they located the equivalent in mice. Given that large sections of the human and mouse genetic blueprints have remained intact since the two species last shared a common ancestor, knowing where a sex-determining gene resided in the mouse helped to trace the corresponding gene in man. Lovell-Badge raises several problems with Sam Berry's Virgin Birth scenario. For example, if Mary had suffered testicular feminization, he believes she would have looked female but would have had testes and been infertile, with little chance of possessing the necessary equipment to give birth: 'Testes produce a factor called AMH, which eliminates the developing female reproductive tract, so such individuals don't have a uterus.' This same point is echoed by Bonthron, who believes that Berry's work relies too heavily on imagination and too little on real-life observations, even of the most bizarre, rare and unusual events. Berry recognizes this problem but maintains that the expression of testicular feminization is variable and 'Mary might have had little or no testis tissue.' However, Robin Lovell-Badge also points out that although parthenogenetic activation of eggs is possible in humans, it usually leads to a teratoma. The highest organism in which parthenogenesis can create a male offspring successfully is the turkey. Ending on this somewhat seasonal note, Lovell-Badge comments, 'That is the closest that we can get.'

three

Santa and Those Reindeer

His eyes how they twinkled! His dimples how merry!
His cheeks were like roses, his nose like a cherry.
His droll little mouth was drawn up like a bow
And the beard on his chin was as white as the snow.
The stump of a pipe he held tight in his teeth,
And the smoke, it encircled his head like a wreath.
He had a broad face and a little round belly
That shook, when he laughed, like a bowl full of jelly.
Clement Clarke Moore, 'A Visit from St Nicholas'

Where do you think Father Christmas is right now? Sitting with a glass of sherry in front of the glowing embers of a log fire in a cosy, wooden house while Arctic snow falls softly on his sleigh outside? Or feeding the reindeer? Perhaps he is poring over his maps, adjusting his route across the North Pole for Christmas Eve? You would be wrong. For the sake of historical accuracy, Christmas cards should show a modern Santa in sunglasses, clad in red and white swimming trunks, sipping a cool Coke next to a swimming pool. For the sake of completeness, a reindeer (called Rosie) with a sunburnt nose should be panting nearby.

There is now evidence to suggest that Santa's grotto lies not in icy Lapland, but among Mediterranean olive groves on Gemiler, a tiny island off Turkey. It is there, historians believe, that St Nicholas, a direct ancestor of Santa Claus, may have died. Gemiler is well known to tourists and has recently been the subject of a number of archaeological studies (latterly by the University of Osaka, Japan, and by a group of scholars including David Price-Williams, an archaeologist who lectures at London University). Though it is only half a mile long, it has at least five churches decorated with frescos and mosaics, and possesses all the hallmarks of a major religious site.

Medieval Venetian sailing instructions refer to Gemiler as the island of San Nicolo. On a church door near the anchorage is a painting of 'Osios Nikolaus' – Nicholas himself. The island also has a huge Byzantine ecclesiastical complex, with a magnificent 300-metre barrel-vaulted processional way. Other Byzantine sites have processional roadways associated with monastic complexes where major saints were venerated, but few ever attained the grandeur of the 'holy city' at Gemiler dedicated to St Nicholas, one of the most popular Christian saints, known today as the patron saint of children, sailors, teachers, students and merchants.

WHO WAS SANTA?

Legend suggests that St Nicholas was born around AD 245 in Patara, an important Byzantine port in Turkey, only a couple of hours' sail from Gemiler. When he was a young man his father died, leaving a great fortune, and Nicholas began anonymously giving away the money to the needy, especially children. Eventually he became Bishop of Myra – the modern-day coastal town of Demre – at the southernmost tip of the Bey Daglari mountains (the name Myra is derived from 'myrrh'). There, he is said to have performed several miracles, including saving sailors from drowning and resurrecting three boys who had been killed by an evil butcher. However, it is through his best-known 'miracle' that St Nicholas evolved into Santa Claus.

A noble and his three daughters had fallen on hard times. The daughters had little chance of marriage as their father could not pay their dowries. They faced a life of prostitution. One night, St Nicholas, hearing of the girls' plight, threw a sack of gold through a window of the nobleman's shabby castle, enough to provide for one daughter's marriage. The next night, he tossed another sack of gold through the window for the second daughter. But, on the third night, the window was closed. Resourcefully, St Nicholas dropped the third sack of gold down the chimney. Townsfolk heard the story and began hanging stockings by the fireplace at night to collect any gold that might fall down their chimneys – hence the tradition of the Christmas stocking and Santa's affinity for fireplaces.

St Nicholas probably died sometime in the middle of the fourth century (one frequently quoted date is 6 December 343). The earliest Byzantine portraits show him with a long white beard, and when the reformed church spread through Europe, he became linked with Christmas because his feast day is 6 December. By the sixth century his fame was widespread, a possible explanation for the huge settlement on Gemiler. But just after 650 this place of veneration was disbanded. The Islamic governor of Syria had launched a fleet to challenge Byzantine sea power in the Mediterranean and quickly destroyed the settlements on Cyprus, followed by those on Rhodes and Cos. Gemiler was abandoned. The site lay forgotten and forlorn – the lost sacred city of St Nicholas.

There are many explanations of how St Nicholas evolved into the Santa Claus we know today. All that is certain is that the modern Santa's roots lie in folk customs and beliefs drawn from a sackload of influences, and that he has appeared in many incarnations. These include the British Father Christmas, the French Père Noël, the Dutch Sinter Klaas, the Danish Jules-Missen and even the Romanian Mos Craicun. The Protestant church strongly influenced the evolution of this icon. Martin Luther objected to the practice of handing gifts to children in the name of a Catholic saint and replaced St Nicholas with a child, the *Christkindlein* (this would mutate, through mispronunciation, into America's Kriss Kringle). In Germany, the *Christkindlein* was joined by a dwarfish, dark-faced companion, often a frightening figure, known variously as Krampus, Pelzebock, Pelznickel (Nicholas in furs), Hans Muff, Bartel or Gumphinkel. There were also female equivalents, Berchtel, Buzebergt and Budelfrau. Most commonly, the companion was called Knecht Ruprecht, and carried a bundle of switches to punish naughty children.

The Dutch are often credited with transforming St Nicholas into the character we know today. Early Dutch settlers in America brought to New Amsterdam (renamed New York when the British took over the colony) their custom of giving presents to children on St Nicholas's Day. In this way, Sinter Klaas, the colloquial Dutch name for Saint Nicholas, evolved into Santa Claus. As described by these early settlers, Sinter Klaas had a broad-brimmed hat, a pipe and short breeches (replacing his original clerical robe). By the beginning of the nineteenth century, the traditions of various countries had started to mingle, so that in 1809, for instance, we see a familiar Santa taking shape in Washington Irving's *History of New York*, in which a jolly, chubby fellow is described riding over the treetops in a wagon.

There is another, quite different, way to trace the evolution of the modern Santa: by representing his development in terms of *memes*, a word coined by the biologist Richard Dawkins to signify the replication of ideas in the human mind through the ages, rather like genes. Loosely speaking, *memes* are units of cultural transmission or patterns of information; they may include tunes, catch phrases, innovative ideas, clothes, fashions and, of course, legends and stories. Santa Claus, Father Christmas and the rest are all examples of *memes*. Genes are carried by organisms in which they produce certain effects (skin colour, blood type and so on) that make each of us unique. Similarly, *memes* are carried by *meme* vehicles – poems, books, computers, sayings and so on – bearing a gift that will distract us or burden our memories. During the frantic run-up to Christmas Day, *memes* can also coerce children to be well behaved. Otherwise, as one particular *meme* makes clear, Santa will not deliver any presents. One sociologist put it this way: 'Parents use the belief in Santa Claus to control children, to induce children to defer demands for gratification until Christmas and to make it appear that Santa, not the parents, causes the deprivation of children.'

MODERN SANTA AND MEANING

It would be a mistake to suggest that today's Santa is merely an amalgam and evolutionary endpoint of his ancestors. For one thing, many of these older versions still exist. In different regions of Germany, St Nick is known by various names, including Klaasbuur, Burklaas, Rauklas, Bullerklaas and Sunnercla. In eastern Germany, where the Santa figure remains more closely connected with his pagan past, he is called Ash Man, Shaggy Goat or Rider. There is also the *Weihnachtsmann* (Father Christmas), a figure who is depicted as tired and stooped from toiling through the dark winter night with his heavy burden of toys.

Anthropologists have set to work on the most ubiquitous form of the modern Santa and declared him to be more than the sum of European influences; indeed, they see him as distinctly American. They highlight five key differences between the Santa of today and his ancestors: Santa lacks the religious baggage of his predecessors; he is, by the standards of Knecht Ruprecht, a bit boring; he has turned into a soft-hearted liberal with no stomach for the punishment meted out by the likes of Sinter Klaas and Knecht Ruprecht; while still a mythical figure, he is more tangible than his predecessors, thanks to appearances in films and TV programmes and in department stores (even in Japan); and he spends much more than his central European forebears, preferring to give Nintendo computer games rather than the more traditional nuts, for example.

So who *is* Santa? The distinguished anthropologist Claude Lévi-Strauss has provided a wonderful pen portrait in the book *Unwrapping Christmas*: 'Father Christmas is dressed in scarlet: he is a king. His white beard, his furs and his boots, the sleigh in which he travels evoke winter. He is called "Father" and he is an old man, thus he incarnates the benevolent form of the authority of the ancients.' Most importantly, says Lévi-Strauss, children believe in him, paying homage with letters and prayers, while adults do not. 'Father Christmas thus first of all expresses the difference in status between little children on the one hand, and adolescents and adults on the other. In this sense he is linked to a vast array of beliefs and practices which anthropologists have studied in many societies to try and understand rites of passage and initiation.'

The sociologists have also attempted to unravel the meaning of Santa. In an article published in 1966 in *The American Sociologist*, Warren Hagstrom of the University of Wisconsin, Madison, ignores the usual definitions given by children ('a fat man with a white beard in a red suit who brings gifts') and plumps for those of the 'sophisticated children' who are his peers. These are couched either in terms of positivism or Clauseology. For the positivist, who recognizes only positive facts and observable phenomena, 'Belief in Santa Claus is defined as erroneous; and the problem of the positivist is to discover how such erroneous beliefs arise. The positivist, arguing that all beliefs arise by inference from experiences, finds the meaning of Santa in false inferences from actual experiences.'

The so-called naturism of Max Müller, the German-born British philologist, is a form of positivism that seeks the origin of figures like Santa in natural phenomena, explains Hagstrom. Children, like primitive men, often personalize the forces of nature: 'While small children may find it difficult to conceptualize the winter solstice, they find it easy to conceptualize Santa Claus. (Ask any child questions about the two phenomena.)'

The Clauseologist position, Hagstrom continues, is that

Santa Claus exists but that his essential nature ('meaning') cannot be empirically ascertained. You must believe in him. 'The empirical phenomena associated with Santa are likely to be illusory and deceptive. It is instead necessary to rely on non-empirical methods of investigation, of which there are two types: inner experience and revealed sources. I cannot report here my inner experiences of Santa Claus, since it has been so long since I've had any genuine experiences of this type ...'

In this piece of sociological whimsy, Hagstrom goes on to say that one of the major problems facing Clauseologists is that of collecting authentic revealed sources. Fortunately, he accepts works like the highly influential 'A Visit from St Nicholas' as part of the canon. This Christmas poem marks perhaps the most important single blueprint for modern Santa. It was written by Clement Clarke Moore (1779–1863), a professor at a seminary in New York. A classical scholar and poet, Moore had translated Juvenal and other Roman poets into English and later turned his hand to poetry in the Romantic style. He was familiar with the folklore of the Dutch, German and Scandinavian immigrants who had settled in the northern United States, including the traditional Dutch tale of Sinter Klaas (whose feast day was by then widely observed on 24–25 December) and the Teutonic and Norse notions of a jovial but somewhat impish figure who presided over the pagan midwinter festivities. In 1822, Moore synthesized these characters into the hero of his poem 'A Visit from St Nicholas'.

That December, Moore read the verses aloud to his children. A visitor to his home was so impressed that he had the poem published in the *Troy Sentinel* in upstate New York. The poem gives us the famous lines: 'Twas the night before Christmas, when all through the house/Not a creature was stirring, not even a mouse.' In dozens of rhyming couplets, often derided today as doggerel, Moore describes a plump, pipe-smoking Santa, who travels from the north in a sleigh drawn by tiny flying reindeer with 'dainty hooves'. This St Nicholas also has a rounded belly ('that shook … like a bowl full of jelly') and a beard that is 'white as the snow'. That much sounds familiar. However, he is described as being 'dressed all in fur, from his head to his foot', which is more reminiscent of Pelznickel than of a latter-day Santa.

The real St Nicholas probably did not celebrate Christmas and probably never saw, or even knew about, reindeer. In Dutch legends, Sinter Klaas travelled on a grey horse and wore bishop's robes. It is not clear whether Clement Clarke Moore had seen a sleigh drawn by reindeer, and it seems most improbable that he had ever seen the beasts in the wild, though he may have known of a Finnish legend concerning 'Old Man Winter', who drove his reindeer down from the mountains, bringing snow with him.

Santa Claus evolved further when he was depicted as a pear-shaped, jolly character with a flowing white beard in drawings by Thomas Nast, published in *Harper's Weekly* between 1863 and 1886. His secularization was by then complete: Nast's Santa was reminiscent of his drawings of a drunken Bacchus and of the corpulent plutocrat William Tweed. Nast himself admitted that he was also inspired by the furs of the Astors when he designed Santa's fur-trimmed clothes.

The world's leading manufacturer of carbonated beverages claims that it created the prototype for the universal twentieth-century red-and-white-clad Santa, even having the cheek to celebrate Santa's sixty-fifth birthday in 1996. According to the Coca-Cola Company, before 1931 Santa Claus took a number of forms in popular culture, from a green elf to a sombre St Nicholas and a gaunt figure calculated to enrage animal lovers by dressing in skins. That year, says Coca-Cola's publicity machine, the company commissioned a young Swedish artist, Haddon Sundblom, to redesign Santa's image.

From then on, Sundblom created at least one Santa picture annually. His St Nicholas wore an ample red coat, trimmed in white and held in place with a thick leather belt, and was depicted in various seasonal settings. In 1934 he was given a hat, also trimmed with white fur. Sundblom removed Santa's pipe, which can be seen in Nast's drawings, but retained his billowing beard, expansive girth and rosy cheeks. Not surprisingly, he also gave him a bottle of Coke. Then came a succession of advertising poses, with children, reindeer, sacks of toys, or letters, but never without that fizzy drink. Santa would gaze intently at the bottle, or grasp it heartily, ready for that 'pause that refreshes'. Of course, without the benefit of a time machine, it is difficult to verify the claims of the Coca-Cola Company.

SANTA: THE HALLUCINOGENIC CONNECTION

There is also a rival theory of the origins of much of Santa's paraphernalia – his red and white colour scheme, those flying reindeer, and so on – which is much more fun, less commercial, more scientific and somehow more appealing (possibly because it is politically incorrect). Patrick Harding of Sheffield University argues that the traditional image of Santa and his flying reindeer owes a great deal to what is probably the most important mushroom in history: fly agaric (*Amanita muscaria*). Before vodka was imported from the east, this was the preferred recreational and ritualistic mind-altering drug in parts of northern Europe.

Each December, this mycologist, or fungi expert, dresses up as Santa and drags a sledge behind him to deliver seasonal lectures on the fly agaric. The costume helps Harding drive home his point, for he believes Santa's robes honour the mushroom's red cap and white dots. Commonly found in northern Europe, North America and New Zealand, fly agaric is fairly poisonous, being a relative of other more lethal mushrooms, the death cap (*Amanita phalloides*) and destroying angel (*Amanita virosa*). The hallucinogenic properties of fly agaric are derived from the chemicals ibotenic acid and muscimol, according to the

International Mycological Institute at Egham, Surrey. Ibotenic acid is present only in fresh mushrooms. When the mushroom is dried, it turns into muscimol, which is ten times more potent. In traditional Lapp societies, the village holy man, or shaman, took his mushrooms dried – with good reason.

The shaman knew how to prepare the mushroom, removing the more potent toxins so that it was safe to eat. During a mushroom-induced trance, he would start to twitch and sweat. He believed that his soul left his body, taking the form of an animal, and flew to the other world to communicate with the spirits, who would, he hoped, help him to deal with pressing problems, such as an outbreak of sickness in the village. With luck, after his hallucinatory flight across the skies, the shaman would return bearing gifts of knowledge from the gods. 'Hence the connotation of the gift of healing, rather than something from the shops, as it is today,' Patrick Harding says.

Santa's jolly 'Ho-ho-ho' may be the euphoric laugh of someone who has indulged in the mushroom. Harding adds that the idea of dropping down chimneys is an echo of the manner in which the shaman would drop into a yurt, an ancient tent-like dwelling made of birch and reindeer hide: 'The "door" and the chimney of the yurt were the same, and the most significant person coming down the chimney would have been a shaman coming to heal the sick.' So how does Harding explain the importance of reindeer in the myth? For one thing, the animals were uncommonly fond of drinking human urine that contained muscimol: 'Reindeer enjoyed getting high on it,' he says. 'Whether they roll on their backs and kick their legs in the air, I am not sure.'

The villagers were also partial to the mind-expanding yellow snow because the muscimol was not greatly diluted – and was probably safer – once it had passed through the shaman. In fact, 'There is evidence,' says Harding, 'of the drug passing through five or six people and still being effective. This is almost certainly the derivation of the British phrase "to get pissed", which has nothing to do with alcohol. It predates inebriation by alcohol by several thousand years.' Such was the intensity of the drug-induced experience that it is hardly surprising that the Christmas legend includes flying reindeer. Witches are said to fly for related reasons. A witch who wanted to 'fly' to a witches' sabbat, or orgiastic ceremony, would anoint a staff with specially prepared oils containing psychoactive matter, probably from toad skins, and then apply it to her vaginal membranes.

More recent records of ritualistic mushroom use also contain references to flying. An Italian saint, St Catherine of Genoa (1447–1510), used fly agaric to soar to the heights of religious ecstasy, according to a study made by Daniele Piomelli while at the Unité de Neurobiologie et Pharmacologie de l'Inserm in Paris. An account of the life of St Catherine describes her taking ground agaric, after which God 'infused such suavity and divine sweetness in her heart that both soul and body were so full as to make her unable to stand'. In Victorian times, travellers to Siberia, Lapland and other regions in northern latitudes returned with intriguing tales of the use of fly agaric by the inhabitants. A mycologist called Mordecai Cooke mentions the recycling of urine rich in muscimol in his *A Plain and Easy Account of British Fungi* (1862). Patrick Harding points out that Cooke was a friend of The Revd Charles Dodgson (Lewis Carroll), the author of *Alice's Adventures in Wonderland* (1865): 'Almost certainly, this is the source of the episode in *Alice* where Alice eats the mushroom, one side of which makes her grow very tall and the other very small. This inability to judge size – macropsia – is one of the effects of fly agaric.'

The toadstool can still be seen in children's books and films – dancing, for example, in Walt Disney's *Fantasia*. In central Europe, clusters of fly agaric are used as Christmas decorations. 'You have all the elements of Christmas in that mushroom,' Harding says, 'the reindeer, the red and white cloak, presents and the chimney.'

RUDOLPH THE RED-NOSED REINDEER

The myth of the reindeer was already well established long before 1949, when the perennially popular singalong-at-Christmas hit 'Rudolph the Red-nosed Reindeer' was released. English texts dating from the Renaissance mention the display of antlers during Christmas dances; this was long before any recorded belief in Father Christmas. Rudolph first appeared only in 1939, in an illustrated booklet by Robert May, which a decade later became the basis for a popular song written by Johnny Marks and performed by Gene Autry, 'The Singing Cowboy'. One commonly held view is that Rudolph's nose was red due to a cold. There is also the implication that while Santa consumed the mince pies left out for him, Rudolph helped himself to the glass of sherry. Perhaps the unexpected triumph of the slightly ridiculous and possibly drunken Rudolph over his sober companions is one aspect of the relaxation of social conventions that has long taken place during winter festivals.

Not so, according to recent research conducted in Norway. Unfortunately for Rudolph, reindeer noses provide a welcoming environment for bugs. They have elaborately folded turbinal bones, covered with blood-rich membranes, which warm the air as they breathe in, and cool it as they breathe out, thereby reducing the loss of both heat and water. (Even when there are icicles and frost on Santa's beard, his faithful reindeer will have dry muzzles.) Odd Halvorsen of the University of Oslo suggested some years ago in the journal *Parasitology Today* that the 'celebrated discoloration' – Rudolph's red nose – is probably due to a parasitic infection of his respiratory system. Even today, he is awed by the response that followed this revelation: 'This paper brought me more fame than anything else I have published,' he says.

Despite living in such chilly conditions, the reindeer share many parasites, including the warble fly, with other ruminants.

Around twenty different parasites are specific to mainland reindeer. *Linguatula arctica*, one of a group of creatures called tongue worms, can be found in reindeer sinuses; larvae of the fly *Cephenemyia trompe* wriggle in the nasal cavity; and nematodes of the genus *Dictyocaulus* squirm in the lungs, as do vast numbers of *Elaphostrongylus rangiferi* larvae. 'We have not been able to quantify the combined effects of these parasites,' says Halvorsen, 'but it is no wonder that poor Rudolph, burdened as he is by parasites, gets a red nose when he is forced to pull along an extra burden like Santa Claus.'

Rudolph notwithstanding, it is something of a puzzle why reindeer in particular are so embedded in modern Christmas culture. They were only one among many kinds of grazing and browsing mammals that once roamed the forest and plains of Europe, northern Asia and North America according to Caroline Pond of the Open University. Indeed, ancient reindeer remains suggest they ranged as far south as Spain and Italy. They had been established for around a million years by the time humans arrived. Now these animals are the most important land-based species for indigenous people in the Arctic. Reindeer meat is delicious, the fur is a good insulator, and the antlers and bones are handy for making tools and ornaments. No wonder the beasts featured in cave art and rock carvings, such as those found at Sagelva, Norway, which date back to 2000 BC.

But the reindeer, says Pond, are badly misrepresented during the Christmas festivities. Take their depiction on Christmas cards, for example. True, reindeer are the only deer species in which both sexes have antlers – formed from bone, often branched, and covered with a thin layer of skin ('velvet') rich in blood vessels – but the males lose their crowning glory around Christmas. The reason has to do with sex. Antlers of mature males are usually larger than those of females, with the most impressive found in caribou and Norwegian reindeer. They probably evolved as a secondary characteristic of males under sexual selection; they depend on the sex hormone testosterone, are larger, more elaborate and heavier in older males and are at their biggest during the breeding season, when they are essential for ritual combat and fighting. The mating season also sees panting, bush thrashing and hunching, when the males urinate on their hind feet. Afterwards they are 'rutted out' (even Rudolph), exhausted by the loss of body weight and fat reserves. No wonder that studies have found that male deer have a shorter life expectancy than the females of the species. At the end of the breeding season, changes in the concentration of sex hormones promote bone reabsorption at the base of the antlers of the adult males. Eventually, the antlers fall off and there is a delay of up to four months before new ones grow in the spring. By inaccurately depicting Rudolph sporting his antlers, Christmas cards are denying the reality of reindeer life.

The Sami (also known as Reindeer Lapps), an ethnic group living in northern parts of Sweden, Norway, Finland and Russia, acknowledge this link between virility and antlers by selling powdered reindeer horn to the Japanese, claiming that it increases potency. The Sami are indeed unusually virile, but the reason, according to a study by Ilpo Huhtaniemi of Finland's Turku University, is not this horny folk medicine but a genetic mutation, found in 40 per cent of Sami men – compared with 25 per cent of other men in Finland and 20 per cent of Swedes – that apparently maintains high-level production of the male sex hormone testosterone in later life.

The fact that Christmas-card artists show Rudolph with a full set of antlers underscores another unfortunate detail, drawn to my attention by Odd Halvorsen: the Sami mostly use castrated male reindeer to pull or carry loads. Without their equipment, males have an abnormal antler cycle so they keep their headgear longer than fully functional males. To keep his antlers for Christmas, Rudolph would have had to be castrated, says Halvorsen.

The more we know about reindeer, the worse are the

problems faced by card illustrators. While the males are squandering their energy on sex and violence, the females are piling on fat, notes Caroline Pond. By the time Christmas arrives, the only adult reindeer with antlers, and enough energy to drag around a sleigh full of presents, are likely to be female.

Reindeer are well adapted to living in a snowy landscape, though one that is more barren than the kind found on Christmas cards. In winter, they dig through snow to feed on the plants underneath. Fine, powdery crystals are easy enough for them to handle but if the snow is too deep or too hard, feeding becomes hard work. Snow that melts and refreezes to form a crust of ice can be so hard that the reindeer cannot dig through it. They don't like the snow to be too deep, or crisp for that matter, says Pond.

Reindeer hide is also misleadingly depicted on Christmas cards. The animals' fur is an efficient insulator. Outer hairs are long and hollow, supporting a fine, dense underfur; together they trap a layer of warm air. Insulation is so effective that snow does not melt on the backs of reindeer. Indeed, Rudolph, Dasher, Prancer and the rest are so well adapted to the freezing cold they would probably find loafing around chimneys and firesides with Santa too warm to be comfortable.

four
Gluttony:
Santa's Genetics

And so I awaited Christmas Eve, and the always exciting advent of fat Santa. Of course, I had never seen a weighted, jangling, belly-swollen giant flop down a chimney and gaily dispense his largesse under a Christmas tree.   Truman Capote, *One Christmas*

Think of the way Santa is usually depicted on Christmas cards. One obvious aspect of his appearance attracts little comment. Not his white beard, ruddy complexion or tendency to 'Ho-ho-ho' at any opportunity. Nor the company he keeps: the reindeer, snowmen and his diminutive helpers. I am thinking of another Santa trait, one that – amazingly for our image-conscious society – is not often discussed. It is that huge stomach.

Generations of children have asked how Santa manages to squeeze down chimneys. But few, if any, have asked the most obvious question of all: why is Santa so fat? After all, if he lost a few pounds, surely his job would be that much easier? Santa would then also provide an influential role model for moderation, restraint and self-control during the festivities. Perhaps Santa's rolls of flesh and overwhelming cheerfulness are the result of seasonal excesses. Perhaps his girth is the result of eating the millions of mince pies left out for him on Christmas Eve.

But what if the jolly, fat Santa is more than just a cliché? What follows is, I admit, speculative. However, recent advances in genetic science suggest that Father Christmas may have a defective gene. Because this gene has been disrupted by a 'spelling error' in his DNA, Santa has a propensity to pile on the pounds. There is even evidence that he may also suffer from diabetes.

Santa is not alone. Obesity is now the most common nutritional disorder in the western world. In America, for example, one-third of adults are overweight. The incidence of obesity – defined as weighing 20 per cent or more above maximum desirable body weight – is rising particularly fast in children. Since 1976, prevalence of paediatric obesity has increased by more than 50 per cent. Eight in ten obese adolescents grow up to be obese adults.

In Britain, one-third of the population carry too much fat and almost 20 per cent are actually obese. The incidence has doubled in the past decade, and worse is to come. A recent report predicted that by 2005, a quarter of British women and one-fifth of men will be obese. In much of Europe, 15–20 per cent of the middle-aged population is classifiable as obese. The picture is better for Scandinavia and the Netherlands, where the figure is around 10 per cent, but worse for eastern Europe, where in some areas it reaches 50 per cent among women. This frequency means that countries such as the UK, France and Germany each have between 5 and 10 million inhabitants who are obese.

The possible consequences of adult obesity include diabetes, high blood pressure, high blood cholesterol, coronary heart disease, sleep apnoea (a life-threatening problem in which breathing stops), gallbladder disease, chronic heartburn, arthritis, certain cancers and depression. Socially, fat people can be as successful as anyone else – think of Santa, for example – but their shape can undermine their self-esteem, particularly in Western societies where thin is fashionable.

Doctors have declared war on obesity. Nowhere is the battle of the bulge more obvious than in America, where obesity leads to an estimated 300,000 deaths each year and costs at least $69 billion per annum in terms of lost working days, health care and other expenditure. Americans spend a further $33 billion a year on weight-reduction products and largely ineffective services offered by the slimming industry. This situation has led scientists to ask why we are getting so fat. As a result, new light has been shed on Santa's stomach.

FAT SCIENCE

Thousands of years ago, fat meant the difference between life and death: being able to store large quantities of energy-dense fuel in the form of adipose (fatty) tissue enabled our ancestors to survive when food was scarce. In the West today, the abundant availability of food and a sedentary lifestyle mean that variations in the way individuals balance energy intake and output make some people more at risk of obesity than others.

When more energy is taken in than is burnt by the body during exercise and while maintaining its so-called basal metabolism (the energy required just to keep it ticking over), the excess is stored as fat. If this continues, obesity will be the eventual result. Studying how we crave, store and use food energy is the key to finding an effective treatment for obesity and perhaps explaining Santa's paunch. Whether you look forward to your Christmas lunch, dread its effects on the waistline, or both, depends on a complicated web of chemical events, from satiety mechanisms at work within the brain to the molecular machinery that lays down deposits of fat in cells.

Hunger pangs, whether experienced by Santa or anyone else, are commonly thought to depend on a range of signals, from the sight of the roast turkey and that wonderful aroma, to the first taste of the meat. The conventional belief is that appetite depends on how the body interprets information from the gut and from hormones in the blood, causing eating habits to be adapted to energy demands. However, since the turn of the century, when it was found that damage to a structure in the brain called the hypothalamus results in obesity, scientists have realized that the urge to eat is caused by something going on in our *heads*. Santa, like the rest of us, is the victim of appetites for food that have been determined by the brain and have evolved in response to the type of diet available during the Stone Age.

Evolution has spent more time optimizing our appetite for the food available on the prehistoric African savannah than for today's fast-food culture. As a consequence, appetites evolved to supply the needs of the body in a tough environment where fat and salt were relatively scarce and famine was never far away. We have inherited a Stone Age appetite for certain nutrients and, as a consequence, it is easier for us to overeat fats, such as ice-cream, than carbohydrates, such as potato, and much easier to overdo carbohydrates than protein. Whereas the protein requirements of primitive humans ranged from 14 to 20 per cent of daily intake, fat appetite was left largely unregulated because immediate energy benefits far outweighed long-term hazards, such as heart disease, which have now become significant as greater numbers of people reach old age.

Aside from those that control our appetites, there are other mechanisms at work processing and storing fat. The first step towards laying down another roll on an expanding midriff takes place in the stomach, where the enzyme pepsin acts on food to break down proteins. When the acidic contents of the stomach enter the small intestine, bile and enzymes are released. They emulsify fats, or triglycerides as biochemists call them. Sugars and protein are similarly broken down, and together these chemical fragments pass into the bloodstream, where they are distributed around the body. Obesity develops when appetite outstrips the body's nutritional needs and excess fat accumulates in specialized fat-storage cells called adipocytes.

A Salk Institute/Harvard Medical School research team, led by Ronald Evans, has discovered that a locally acting signalling chemical, called a prostaglandin, tells cells when to store fat. The prostaglandin sends this message by docking with a receptor, a specialized protein found within the nucleus of cells. This union acts like a switch, turning on genes located in the nucleus and instructing the cell to develop into a fat cell. An enzyme called lipase is then secreted into the bloodstream by the fat cells, breaking down the fat (triglycerides) into its basic chemical building blocks, fatty acids and glycerol. In this form, they can be taken up by the cells and then reconverted into triglycerides to create fat tissue. A second type of lipase is present in the cell which the body can summon to break down the stored fat and convert it into energy.

Recently, another part of the fat puzzle was fitted into place. A second receptor found in the liver appears to function as a sentinel that monitors fat consumption. Eat a slice of pizza and the sentinel sounds the alarm. The body responds by triggering a fat-burning pathway. This discovery casts new light on the effects of fat on the body. It turns out that the sentinel is not able to sense all fats and some go undetected. Ronald Evans points out that, interestingly, polyunsaturated fats are easily noticed by the sentinel and thus might be considered 'good fats' because they promote their own removal. By contrast, saturated fats are less easily detected and would therefore be considered 'bad fats'. Thus, what Santa eats is just as important as what he doesn't eat.

Scientists have also discovered that certain genetic defects are linked to obesity, and for this reason, some of us are more likely to indulge in a Christmas blow-out than others. Some of these genes carry faults that mean cravings for fatty food go on long after the body's capacity to burn these calories has been exhausted. Other defects interfere with the process by which fat is laid down in the body. Every molecular clue presents a new target for drug designers seeking the grail of the anti-fat pill – and another hint why Santa is so plump.

FAT MICE

Our understanding of obesity owes a great deal to the breeding of special strains of fat mice and to efforts by scientists to track down the genes that make them overeat. You may wonder what podgy rodents have to do with humans, let alone Santa. First, they breed quickly and you can do things to them you would never dream of doing to a human, let alone Santa. Second, it turns out that if a gene is significant in mice, one very similar is probably also at work in humans. Scientists say that nature tends to be highly conservative. This stands to reason, since we all evolved from common ancestors, which in turn evolved from a primitive bug that lived 3.85 billion years ago. If a piece of molecular machinery that controls cell division works for yeast, for example, why bother to change it for humans? To date, a handful of genes have been discovered in mice that influence obesity. Known as *tubby, fatty, diabetes, obesity, mahogany* and *agouti*, each gene plays a role in maintaining a particular amount of stored energy, in the form of fat, and each is likely to be present in humans too.

The quest to uncover the link between genetics and obesity can be traced back to 1950, when a team at the Jackson Laboratory in Bar Harbour, Maine, discovered a fat strain of rodent they dubbed the 'obese' (*ob*) mouse. It was not until 1994, however, that the reason why these rodents were so fat became clear – it was down to a defect in a single gene, also named *ob*. The discovery was made by six scientists, including Jeffrey Friedman and Stephen Burley, from the Howard Hughes Medical Institute and the Rockefeller University in New York. Their findings were published in the journal *Nature*; the cover of that issue showed weighing scales in which two slim mice were outweighed by one grotesquely podgy colleague, looking like a furry ball. This mouse suffered from a fault in the *ob* gene. This gene is the blueprint for a protein called leptin, a hormone that tells the brain how much fat is deposited in the body and is thus part of the 'stop-eating' message.

The researchers measured the amounts of leptin in the blood of both normal and obese mice, and found that fat mice with a defective *ob* gene do not make the hormone. Further evidence pointing to the importance of leptin came when Friedman and his team found it to be present in the blood of six lean humans. 'Our findings indicate that when *ob* is defective, leptin is not made and does not transmit its signal to stop eating,' reported Jeffrey Halaas of the Rockefeller University. Consequently, mice with a faulty *ob* gene are overweight. Injections of the missing protein cut the body weight of these mice by 30 per cent after two weeks of treatment.

Jeffrey Friedman likened the body's fat-control mechanism to a thermostat: the 'fatstat' senses leptin levels, which act as a chemical barometer of the amount of body fat, and adjusts fat deposits accordingly, by controlling appetite, thus regulating how much food is eaten and how much is turned into energy in the body. Meanwhile, other work has shown that feelings of hunger are stimulated by the action of another molecule in the brain, called neuropeptide Y (NPY), which acts on 'feeding receptors' in the hypothalamus. Subsequent research showed that when NPY is genetically removed from the obese mice, the mice lose some weight.

Although leptin helps the brain to weigh up fat deposits in the body and NPY stimulates appetite, these are just two players in what Friedman has called the alphabet of weight control. For example, although high levels of NPY can increase food intake, the molecule is probably not involved in normal eating. It acts by docking with a receptor called Y5, and when that receptor is disabled in a mouse, the rodent retains a normal appetite, suggesting that NPY is not critical in telling animals when to eat.

Although we have yet to understand the details of how this alphabet spells out FAT, the original discovery of the *ob* gene and leptin caused a sensation. Injections of leptin can fool the body into thinking it carries too much fat, reducing appetite and increasing energy consumption. The biotechnology company Amgen, which is based in Thousand Oaks, California, reportedly agreed to pay a $13 million signing fee to Rockefeller for rights to the gene.

What is true for mice does not necessarily apply to humans, or to Santa for that matter. However, since the pioneering work at the Rockefeller University, the human equivalent of *ob* mice have been reported by Sadaf Farooqi and Stephen O'Rahilly of the Addenbrooke's Hospital in Cambridge, England. Cousins aged two and eight, born in Britain of Pakistani parents, provided the first proof that leptin plays a crucial role in weight control; this discovery was an important landmark in human obesity research. The eight-year-old weighed 89kg (14 stone) and had already had liposuction on her legs to help her to move around. The two-year-old weighed 28.5kg (4½ stone). More than half his body weight was fat. Both children ate non-stop from the time they were babies as a result of low levels of leptin. This not only pointed to an obvious treatment – once-daily leptin injections, which are showing significant benefit in terms of appetite suppression and weight loss – but also raised the possibility that even people who are mildly obese may carry variants of this genetic defect.

So could Santa himself suffer from a defective leptin gene? Yes. Stephen Burley, Jeffrey Friedman's colleague, believes that Santa's pendulous tummy is likely to lead to adult-onset diabetes – the most common type – because excessive fat promotes resistance to the insulin hormone that regulates blood-sugar levels. The pancreas begins to make more and more insulin to keep up but eventually runs out of steam so that there is not enough. The result is diabetes, when sugar metabolism goes awry.

Ob is not the only defective gene that might be responsible for Santa's girth. The action of another, called *agouti*, was reported in January 1997 by a team at Oregon Health Sciences University. Faults in this particular gene cause a form of obesity in mice that is even closer to human obesity than that associated with defects in the *ob* gene, making animals 20–50 per cent heavier than normal. 'The mice develop a moderate form of obesity rather than the morbid kind of obesity induced by some of the other mouse genes associated with obesity that have been discovered to date,' says Roger Cone, senior author of the study. 'Most humans who are overweight are moderately heavy, not morbidly so.'

The *agouti* gene was so named because it has long been known to cause a yellow coat colour in mice and other animals by preventing the production of dark pigment, called melanin, in hair follicles. It does this by interfering with a key molecular docking site – a receptor – on the surface of pigment cells. This site is called the MSH receptor and is responsible for stimulating the production of the melanin pigment. The protein made by the *agouti* gene blocks the receptor and thus the production of the dark pigment, which allows yellow pigments (phaeomelanin) that are already present to be readily seen.

The protein made by the *agouti* gene, a mutated version of a naturally occurring gene, also blocks an MSH receptor in the brain cells, interfering with a signalling pathway in the hypothalamus, a region of the brain involved in feeding. Roger Cone explains: 'When mutated forms of the *agouti* gene are inappropriately expressed in the brain, normal feeding behaviour is disrupted and the mice become obese and develop early symptoms of diabetes.' The animal overeats because the mutant *agouti* gene interferes with a signalling pathway involving a neuropeptide (that is, a small stretch of protein that carries information in the brain from one neuron to the next) called melanocortin. In a normal mouse, when melanocortin binds to a particular kind of MSH receptor it exerts an inhibitory effect on eating and causes the animal to stop feeding, so these mice do not get fat.

If Santa had a mutated *agouti* gene he would fail to receive an inhibitory signal from the brain, and would consume excessive amounts of mince pies, turkey and other seasonal food. This hypothesis has been backed by a study conducted by Millennium Pharmaceuticals in Cambridge, Massachusetts, in association with Roger Cone's team. The researchers concluded that specially bred mice lacking the key receptor overate in exactly the same way as *agouti* mutants.

There are many more genes in the genetic alphabet soup of pathways that influence feeding. Another candidate gene to explain Santa's paunch was found by researchers at the University of California Davis Medical Center, the Duke University Medical Center in North Carolina, and the Centre National de la Recherche Scientifique outside Paris. Duke's Richard Surwit believes that the gene may help explain why some people can eat a rich diet and stay slim while others eat the same diet and get fat: 'We believe this is at the heart of what happens in people who get fat.'

Given the somewhat dull name of uncoupling protein 2 or UCP2, the gene contains the blueprint for a previously unknown heat-generating protein that exerts an effect on the energy expenditure of the body. This takes three forms: physical activity, resting metabolic rate, and thermogenesis or heat production. UCP2 affects the last of these processes by burning calories as heat rather than storing them as fat. As a result, people who have more of the protein burn more fat, while people who have less of the protein store more calories as fat. Even small changes in heat production could potentially have a big impact on a person's weight in the long term. For a typical adult, just a 1 per cent reduction in body heat could translate into a 2.3kg (5lb) weight gain over the course of a year, all other things being equal. By the same token, raising body temperature by just 1 per cent would result in a 2.3kg (5lb) weight loss. Obesity could thus be a problem caused by a one-tenth of one degree alteration in body temperature. The average healthy human body temperature is 37°C (98°F), but an individual's normal temperature may be one or two degrees higher or lower than that.

The scientists think of the UCP2 gene as a missing link that helps us understand the connection between body temperature and weight. While the discovery will not solve all weight problems, it could lead to simple therapies for some overweight people to increase their 'expression' (use in the body) of the gene so that they burn off more of the calories they eat as body heat rather than storing those calories as fat. This discovery makes it theoretically possible to use the gene only in targeted tissues to reduce the size of fat deposits in certain areas of the body, an idea Richard Surwit calls 'genetic body shaping'.

RESHAPING SANTA

Remodelling Santa's body does sound rather excessive. However, scientists have already demonstrated in their favourite experimental subjects that it is possible, through genetic engineering, to produce a creature that can eat huge amounts of fatty food yet never get fat. Stanley McKnight's team at the University of Washington School of Medicine in Seattle bred mice that lacked a gene – called RII beta – involved in the complicated and delicate balance of metabolism. RII beta codes for a segment of an enzyme called PKA, which regulates fat storage and metabolism. Knocking out the gene made the mice more sensitive to hormones that break down white fat, the kind responsible for spreading midriffs. But the more important part of the process takes place in a related kind of tissue, known as brown fat. This is the body's generator, converting fat into heat. It is so called because the cells have extra mitochondria, the power packs of cells, which tint the tissue brown.

In mice who have lost the RII-beta gene, the PKA enzyme is overactive, stimulating fat all over the body to burn off energy. The brown fat kicks into high gear, using stored white fat for fuel. 'Brown fat is acting like a little furnace,' Stanley McKnight says. The result is perpetually slender animals. These mutant mice were found to store about 50 per cent less fat than their normal counterparts when fed the typical low-fat laboratory food. When they were given fatty lab fare, the mouse-food equivalent of the French fries, hamburgers and ice-cream consumed by many in the West, the difference became even more pronounced. The mutant mice had about half of the fat of normal mice, although they did not have fewer fat cells. They had a higher metabolic rate and a slightly higher body temperature. 'The more food and calories they take in, the more obvious it is that they won't get obese,' McKnight observes. 'They're protected from obesity.'

Other labs have developed genetically skinny mice, but most of these have had physical problems related to the genes that were knocked out. McKnight's mutants, however, appear to be normal, living as long as other mice and maintaining fertility. Using genetic engineering, it may even be possible one day to interfere with the PKA enzyme itself to ensure that we are all fast burners, those rare and infuriating people who can eat all the fried foods they want and never gain weight. Another approach opened up recently with the discovery that daily doses of a protein called gAcrp30, which causes muscle to burn fatty acids faster, may allow people to gorge themselves on a high fat diet yet still lose weight. And then, when it is possible to gorge on festive fare without risking an expanding waistline, it will be fascinating to see whether the popular image of Santa will evolve into a more svelte Kriss Kringle.

WHY SANTA LAUGHS IN THE FACE OF OLD AGE

Two other Santa characteristics deserve scientific comment. First, his jolly nature. This could be a result of his genetic propensity to obesity. Plump, bouncing baby boys are more likely to grow into happy men, while skinny babies often get the blues in later life, according to a study by Ian Rodin of Southampton University, involving men born between 1911 and 1930 in Hertfordshire, England. Rodin explains how factors during pregnancy and the first years of a baby's life could alter brain chemistry and hormonal responses, determining whether an individual has a sunny or gloomy disposition in later life. Other work in Southampton on the same data has shown a link between birth in a cold climate – such as that found at the North Pole – and adult obesity.

Another puzzling thing about Santa is that, though elderly, he never ages. He seems to have found a way to arrest the ageing process so that, as the years roll past, he remains sprightly enough to clamber down millions of chimneys. There is a school of scientific thought that says lifespan is determined by a fixed amount of metabolic activity: eat less, slow your metabolism and you may live longer. Experiments on rodents and monkeys provide powerful support for this idea. However, Santa's expansive waistline rules out this particular explanation. Age at death is at least partly under genetic control, and the race is on to identify the genes and molecular mechanisms that are responsible so that they can be manipulated. Scientists and pharmaceutical companies should look no further than Santa. There is good evidence that he may have beaten them to the prize of a Methuselah pill.

We can speculate about how Santa stays so old, but no older. One idea is that he has developed a method to manipulate structures called telomeres that are found on the ends of chromosomes and stop them from 'fraying', rather like the plastic bits at the ends of shoelaces. Telomeres shorten with repeated cell division until, when they have dwindled to a certain point, the cell can no longer divide and becomes senescent. Santa may have found a way to prevent telomeres from shortening, perhaps by activating an enzyme called telomerase. The only problem with this theory, however, is that the cells of even the oldest people still have quite a few more divisions to go, suggesting that there is more to the anti-ageing story than this.

Studies involving a little worm called *Caenorhabditis elegans*, a nematode, have suggested that one factor contributing to Santa's longevity might be his lack of a partner. If the experience of the worms is anything to go by, men could add years to their lives if they spent less time and energy pursuing the opposite sex, says David Gems of University College London. His suggestion dovetails with a study of humans by James Hamilton and Gordon Mestler, published in 1969 in the *Journal of Gerontology*, which showed that eunuchs lived on average 13½ years longer than intact males. However, despite this tantalizing clue to the secret of Santa's longevity, Gems says that there is 'really no evidence either way that the level of sexual activity per se shortens human life span.'

Another worm-inspired approach in the quest for longevity is to tinker with genes that control the biological timetable from the cradle to the grave. The first-ever life-extension mutation, in a gene called *age-1*, was found earlier last decade by Thomas Johnson at the University of Colorado. Other genetic mutations have since been discovered; one of these occurs in the gene *daf-2*. Work by the 'worm group' at the University of San Francisco, California, has shown that *age-1* and *daf-2* mutations act in the same lifespan pathway and extend lifespan by triggering similar, if not identical, processes. An important clue to what is going on in these aged worms came from the work of Gary Ruvkun at Massachusetts General Hospital, indicating that *daf-2* may regulate glucose (sugar) metabolism.

According to Ruvkun, when *daf-2* is defective, the worms are unable to respond to 'worm insulin' and enter a state of hibernation, usually triggered by starvation. However, those possessing only a slight defect in the gene do not hibernate but shift their metabolism so that they can live longer. Humans have a similar gene that controls the way in which the body reacts to insulin, the hormone that is secreted in response to a rise in blood-sugar levels, causing numerous changes in the metabolism of cells. This should be no surprise. Genes of key importance tend to do the same thing, whether in worms or men, as the first analysis of the human genetic code in 2001 confirmed.

The discovery by Gary Ruvkun that insulin regulates the life span of worms suggests that the rate at which humans age may be closely connected to how we burn the calories we eat. Backing for this idea has come from a mutation in a gene, one found in humans, that doubled the age of fruit flies so that they lived the equivalent of a human life span of 150 years. Discovered by Stephen Helfand and colleagues at the University of Connecticut Health Centre, it was named *Indy* in homage to the film *Monty Python and The Holy Grail* in which a plague victim uttered the words 'I'm not dead yet' while being hauled off for burial.

Indy may affect ageing by creating a metabolic state similar to that of being on a diet. Mutations in *Indy* may either mean that calories are not absorbed, or are wasted. Helfand said: 'It would be as if the *Indy* animal can eat as much as it wants without becoming obese, live twice as long as average, and still retain normal function and activity.' In both humans and flies, the gene is found where the body stores energy and uses it. *Indy* absorbs essential nutrients through the gut, concentrates them in the liver, and reabsorbs them via the kidney.

Perhaps Santa has found a way to tinker with his cellular glucose machinery to slow his ageing process. There are, however, potential pitfalls facing any wannabe gene-tinkerer. Although some long-lived worms seem to lead normal lives, other gene mutations can cause the nematodes to live in slow motion. One such was found at McGill University in Montreal by Siegfried Hekimi and his colleagues, who created nematodes that lived up to 50 per cent longer than normal. This particular ageing gene does not seem to correspond to what is known about our Santa: he has a frantic workload on Christmas Eve that would have discouraged him from fiddling with the slo-mo gene. Nor have I ever heard it claimed that he has 'Hoooo-hoooo-hooooed' rather than 'Ho-ho-hoed'.

This work on flies and worms is all very well, but what about humans and what about Santa? In 1996 scientists found the first human gene known to affect the ageing process, one responsible for a rare disease called Werner's syndrome. Sufferers appear to age in adolescence. They go grey, develop wrinkled skin, then lose their hair and become susceptible to disease associated with old age, such as cataracts, heart disease, diabetes and cancer, usually dying before the age of fifty. Darwin Molecular Corporation, a Seattle biotechnology company, and a team at the Seattle Veterans Administration Medical Research Center, led by Gerard Schellenberg, showed that these people carry a faulty version of a gene associated with a type of enzyme known as a helicase. These enzymes split apart, or unwind, the two strands of the DNA double helix, a process that must take place in healthy dividing cells if they are to pass on their genetic material, in the form of chromosomes, to daughter cells. If a lack of helicase accelerates ageing, then introducing extra copies of the gene into the body can cause a person to live longer. More recently, a team developed *Methuselah* mice in the first experiment to suggest that mammals, including humans, can live longer if the body's resistance to chemical damage is boosted. Earlier work had revealed a link between increased lifespan and a robust response to oxidative stress − damage by reactive chemicals in the body, notably forms of oxygen. Pier Pelicci of the European Institute of Oncology, in Milan, and colleagues found a mouse with a defect in a gene responsible for a protein called p66shc lived a third longer than mice without this defect. Experiments showed that its cells resisted damage by hydrogen peroxide and UV light, confirming that ageing involves oxidative damage. Perhaps the protein triggers the death of wrecked cells, or must be harmed by oxidation for repair mechanisms to be turned on. Whatever the explanation, Pelicci and his colleagues say that a better understanding of these processes will pave the way for better understanding of ageing in mammals, including humans. No doubt many more opportunities to arrest ageing will emerge from understanding such fundamental processes. By tinkering with them to stay a sprightly sixty-something, Santa can ensure that he will be able to deliver presents for many years to come.

five

The Flame and the Tree

> I have been looking on, this evening, at a merry company of children
> assembled round that pretty German toy, a Christmas Tree. The tree
> was planted in the middle of a great round table, and towered high
> above their heads. It was brilliantly lighted by a multitude of little
> tapers; and everywhere sparkled and glittered with bright objects.
>
> Charles Dickens, *A Christmas Tree*

A fir tree festooned with flickering candles is one of the arche-
typal Christmas-card images. The evergreen and the candle
celebrate the same thing – life-giving sunlight – and are ancient
symbols dating back long before Prince Albert introduced the
tree to Britain, or indeed, before Martin Luther supposedly
bedecked a tree with candles to remind children of the heavens
from which Christ descended to save us.

Our ancestors held winter festivities to usher in the annual
return of sunlight, warmth and fertility with rituals involving
the yellow light of a living flame that seemed to defy the cold
winter months. Today, Christmas and other seasonal celebrations, such as Kwanzaa and Hanukkah, are united by this
symbol of the rebirth of the sun's life-giving energy.

We can gaze deeply into the workings of the living world by
studying what happens when we light a candle at Christmas-
time. The flame marks the last step in an extraordinary series of
physical and chemical processes that first capture sunlight to
forge chemical bonds in wick and wax, then snap these bonds
to release the stored light. The first and most important link in

the chain is photosynthesis (from 'photo', meaning 'light', and 'synthesis', meaning 'building something'). This process drives the living economy that thrives on the surface of our planet. Each year, green plants such as Christmas trees harness the energy of sunlight to pluck 100 trillion kilograms of carbon dioxide from the skies, then combine it with hydrogen from water to build their food – carbohydrates – and release oxygen. One acre of Christmas trees can produce the daily oxygen requirement of eighteen people. The United States has approximately one million acres of growing Christmas trees; that means that around 18 million people each day are supplied with the oxygen generated as the trees harvest sunlight.

What is so beautiful about the act of lighting a candle on a Christmas tree is that it celebrates the cycles that turn within and without living things. The chemical energy generated by photosynthesis in plants is passed up the food chain, for instance to grazing cattle, and thence on to a tallow candle. When it is lit at the gloomiest time of year, the candle releases this energy and returns the complex fat, or wax, molecules to the form in which the plants found them – water and a hot breath of carbon dioxide that can again be incorporated into living things.

THE CHRISTMAS TREE

Like so many aspects of the Christmas celebrations, the origins of the tree symbol stretch back to prehistoric times. Ancient people were fascinated by how some trees and plants continue to thrive among the dead branches of a forest in winter. To a primitive mind in a deciduous world, an evergreen would probably have suggested permanence and a magical ability to flourish with little help from the sun. The ancient Egyptians brought green palm branches into their homes on the shortest day of the year as a symbol of life's triumph over death. The same symbolism is also apparent in the Roman festival of Saturnalia, when buildings were decorated with evergreen branches of holly, pine and ivy in honour of Saturnus, the god of agriculture. Holly, ivy and mistletoe are not only green but also bear recognizable fruit during the winter, again cocking a snook at the elusive sun and keeping alive the hope that a fruitful year is to come.

This triumph of fertility over the elements is also echoed in the English legend of the Glastonbury Thorn, supposedly planted by Joseph of Arimathea. The legend goes that soon after the death of Christ, Joseph came to Britain to spread the message of Christianity. Being tired from his journey, he lay down to rest and pushed his staff into the ground beside him. When he awoke, he found that the staff had taken root. The resultant bush was the Glastonbury Thorn, which flowered each year on Christmas Day.

WHY CHRISTMAS TREES ARE EVERGREEN

An explanation of the processes that cause trees to shed or retain their leaves in winter adds an entirely new dimension to the ancient obsession with evergreens and their defiance of the weak winter sun. Botanists have long asked why some plants are evergreens. Leaves and needles are highly efficient 'solar cells' and to understand why a Christmas tree hangs on to this power supply during the dark winter months, while other trees discard it, we need to know a little more about the biochemical machinery that turns within leaves. These needles, fronds and palms of green tissue provide plants with energy by harnessing sunlight using photosynthesis. This occurs inside structures called chloroplasts, contained within leaf cells where pigment molecules capture light energy. Chlorophyll is the most important pigment, absorbing red and blue light, and allowing green wavelengths to reach the eye. The light energy, once captured, is used to convert water and carbon dioxide from the atmosphere into oxygen, which is released into the air, and glucose, the food used by plants.

The winter sees less daylight, combined with a sun that hangs lower in the sky. That means less photosynthesis and thus less energy to sustain the tree. Moreover, temperatures are cold enough to threaten cell damage. Trees have adapted to this seasonal change by resting throughout the period and living off excess food stored during the summer months. This process is not solely triggered by cold weather, even though chemical reactions within the leaf do slow in response to lower temperatures. The signal for dormancy is the decline in the hours of sunlight, and the greater periods spent in darkness. Trees have evolved different strategies to deal with the annual decline in sunlight that takes place during winter; these can be categorized as either deciduous (Latin for 'falling') or evergreen. Both types of tree stop growing in winter, but whereas deciduous trees shed all their leaves during this period, evergreens such as conifers do not. (They shed their needles, however, for example when deprived of water or bathed in shadow, but not all at one particular time.)

One might think that there is a simple reason to explain why deciduous trees are stripped during winter: their broad leaves are less able to cope with winds and other vagaries of winter weather than the more aerodynamic needles of a fir. To some extent this is true. Nonetheless, the loss of broad leaves is deliberate: it may be hastened by breezes and blizzards but it is not dependent on them. According to Bill Proebsting of Oregon State University, an elaborate cellular mechanism governs the process by which leaves and deciduous trees part company. At the base of each leaf is a special layer of cells called the abscission zone. When the time comes to shed a leaf, the cells in this layer begin to swell, slowing the transport of materials between the leaf and the tree. Once the abscission zone has been blocked, the mortar between its cells is set to dissolve. A signal is sent to the zone, where enzymes break down the matrix holding the cells together. A tear line is formed at the top of the zone and progresses downwards, until eventually the leaf is blown away or falls off. Once the leaf falls, the stem side of the abscission zone forms a protective layer to seal the wound, preventing water from evaporating and bugs from getting in.

The question might be asked, then, why some plants bother to be evergreen at all. A range of factors has influenced the evolution of Christmas trees and other conifers, causing them to hang on to their needles in the winter rather than shed them all in one go. According to Peter Davies, a British scientist who is now a professor of plant physiology at Cornell University, conifers and holly retain their leaves to take advantage of occasional winter sunshine. Unlike deciduous leaves, those of conifers must be coated with resins or wax (which prevent moisture loss) and also possess modified biochemical machinery, which prevents cell damage from the winter cold. But this also means that evergreens require more energy to make and maintain leaves compared with deciduous trees, which have leaves that are too delicate to withstand low temperatures. The latter group of plants has decided, in effect, to cut its losses and not invest any more resources in its solar panels.

Within each leaf on a deciduous tree, the green chlorophyll that captures sunlight and the biochemical machinery that converts sunlight into food for the plant gradually degrade. As the green pigment fades it reveals other pigments, notably the yellow and orange carotenoids that provide vivid autumn colours, and reduces the ability of the leaf to harness sunlight. By shedding leaves, deciduous trees can maintain overall efficiency at a slightly higher level. Once the plant has withdrawn any useful nutrients and stored them in the cells of the trunk for use the next spring, leaf loss can also help to rid it of the by-products that build up in leaves. One suggestion is that leaves are 'excretophores' and that the shedding of a yellowed leaf is the equivalent of going to the lavatory.

The use of the fir as a Christmas tree is thought to have started in ancient times in the Black Forest in south-west Germany, while evergreens and small trees were part of the winter-solstice festival of pagan tribes. It is not known when the fir (*Tannenbaum* in German) began to be adopted in other regions. Another component in the festivities is the yule log, supposedly dating back to the logs of oak burnt by Norsemen in honour of the god Thor, which were big enough to burn throughout the longest winter nights and were thought to help usher in the return of the sun. In addition, the resulting ashes were said to have the power to protect a house against lightning, to cure disease and to fertilize fields.

Legend has it that Martin Luther (1483–1546) was the first person to decorate a tree. He was so moved by the brightness of the millions of stars on a winter's night that he set candles on his tree to simulate the effect. In fact, the custom of decorating trees dates from much earlier, and can be traced to tree-dressing rituals practised from Russia to India. Celebrated mythological trees have included Yggdrasil, the Nordic tree of life; the Indian Bodhi tree; and Eden's tree of knowledge. In some pagan rituals a tree was decorated to encourage the tree spirits to return to the forest so that it would sprout again, which of course it did every spring.

Decorated trees are recorded in 1605 in homes in Strasbourg, then a part of Germany. By 1796 we have the first picture of a candlelit Christmas tree, an illustration showing Christmas Eve at Wandsbek Castle near Hamburg. The Christmas tree was eventually exported to the United States by German settlers and by Hessian mercenaries paid by the British to fight in the Revolutionary War. In 1804, two hundred years after the first sighting of decorated trees, soldiers stationed at Fort Dearborn (now Chicago) were seen dragging trees from the surrounding woods to their barracks at Christmas. By 1842 the custom had reached Williamsburg, Virginia, and almost a decade later, two ox sleds loaded with trees were hauled from the Catskills to the streets of New York to become the first retail lot sold in the United States. By the end of the nineteenth century, the tree was so much part of the festivities that a description of Christmas at the White House refers to 'an old-fashioned Christmas tree'.

In Britain, the custom of decorating Christmas trees was known to the socially aware from the middle of the eighteenth century because of the German background of the Hanoverian monarchs. However, it was not until Queen Victoria and Prince Albert set up a Christmas tree for the first time at Windsor Castle in 1840 that the symbol was firmly fixed at the heart of Christmas. Eight years later, Victoria and her family appeared beside the tree in the *Illustrated London News*, and the custom became widespread in homes throughout Britain.

Today, the symbolism of the flame and the tree burns as brightly as ever. Around 40 million American families celebrate the holidays by putting up a freshly cut or living tree. Tens of millions more trees are decorated worldwide, from banana trees dressed by Christians in India, to the trees decorated with glowing tapers planted on graves in Oberammergau, Germany, where the dead are included in the Christmas celebrations. But the traditional Christmas tree is under threat from a vulgar impostor: plastic trees that shed no needles, show no flaws, and can be used year after year. This is taking the deep-rooted obsession with the evergreen a step too far.

SUPERTREE

Thanks to the technology of cloning, however, the real thing is staging a comeback, and teams of scientists are developing methods of mass-producing thousands of copies of an individual high-quality Christmas tree. The problem with the old-fashioned varieties is that their seeds contain genetic variability, not only spelling imperfection for the consumer but also creating a headache for the grower when it comes to harvesting trees of many shapes, girths and heights. To tackle this problem by producing genetically homogeneous tree seed in the traditional way would take about seven generations of breeding.

With fir trees, each generation takes fifteen to twenty years to reach maturity, so it is just not commercially feasible to obtain genetically uniform seed this way. The alternative is to use cloning to create in-bred lines of trees.

Dan Keathley's team at Michigan State University's Department of Forestry is focusing its efforts on cloning true firs, Douglas firs and Scotch pine. The first step, says Keathley, is to find 'individuals that are really select – what we are talking about is the one tree in 10,000.' This flawless tree should have the following properties: a straight trunk that easily slips into a stand; the strength to hold lots of ornaments and tinsel; thick needles; good colour; limbs angling upwards at 45 degrees; a uniform conical shape tapering upwards at 35–45 degrees; and good needle retention. Growers want these properties to be coupled with rapid growth and disease and insect resistance. Mass production of such super-trees should cut costs.

Once the super-tree has been selected, a piece is cloned to produce a field of identical trees by one of two basic methods. In one process – the tissue-culture technique – the researchers take a bud or seed from a tree, sterilize it to kill fungi or bacteria, and place it in a nutrient medium that promotes growth into a callus, a clump of randomly dividing cells. The callus can be broken into many pieces, each of which puts out shoots, which are then placed in a medium that causes roots to grow using hormones. From there each shoot becomes a whole tree again.

Another method – micropropagation – relies on a bud taken from a parent tree. Shoots are produced with the help of hormones and are then cultivated to form a tree. 'The best approach is the bud method,' said Dan Keathley. This avoids going through unorganized callus tissue, where chromosomal problems arise. 'We are working to develop orchards for all of the commercially important Christmas-tree species used in Michigan.'

Hundreds of cloned Virginia pines have already been grown

by another team at the Texas Agricultural Experiment Station. 'The cloned tree is the future for Christmas trees in the South,' says Don Kachtik, a grower whose test-tube-to-tree-farm pines are among the first in the state ready to be cut for the holidays. 'We only have one type of tree that grows well in south-east Texas, the Virginia pine, so we need a good one.' Craig McKinley, who with Ron Newton selected good 'parent' trees for the experiment and developed the cloning technique, explains further: 'When we find a tree with good genetic material, we can capture improvements faster than going through the usual propagation.' The traditional Virginia pine seedling costs less than 7 cents compared to the current cost of 75 cents to $1.50 for a cloned tree ready to plant. However, according to Newton, once they have become a common feature of the festivities it is unlikely that the cloned trees, especially those improved genetically to resist insects and disease, will cost consumers more. Mike Walterscheidt, forest specialist for the Texas Agricultural Extension Service, agrees: 'Growers hope that with cloning, they will be able to sell 95 per cent of the trees they plant. Right now, only 60 per cent to 70 per cent are sold because the others are ugly or they die from disease or something.'

Another method of creating a super-tree is by genetic engineering, in which genes are inserted into the tree to introduce novel and desirable characteristics. Interest in engineering so-called transgenic trees has come from the wood-pulp industry, who see this as a way to make cheaper paper products, not least Christmas cards. Pine and spruce fungal diseases cause significant losses in tree plantations and finding a gene resistant to them is one of the genetic engineers' main aims. Resistance to insects is another major criterion: Christmas-tree plantations are prone to attack by soil-borne insect larvae such as white grubs. And by manipulating genes that trigger senescence, the deterioration that accompanies ageing, it may be possible to slow down needle loss. Krishna Podila of the Michigan Technological

University believes there are many reasons to engineer a super-tree, such as speeding up the growth rate: 'If you can introduce genes that would allow the trees to grow faster, you can reduce the turn-around time on these plantations,' he says. Podila's laboratory was one of the first to conduct genetic engineering on a conifer – a larch – and some progress has also been reported by colleagues in Madison, Wisconsin, and in New Zealand in generating genetically engineered spruce and radiata pine, respectively, although conifer trees are proving difficult to manipulate in this way.

Those who fear that one day a 'Frankentree' may run amok need not worry, or so it is claimed. For one thing, trees containing value-added genes to promote fast growth and resistance to herbicides and disease will also be designed to be sterile, thus protecting the environment and the interests of manufacturers of paper products by removing the possibility of 'genetic pollution'. At the moment, however, the commercially produced supertree is but a glint in the eye of the biotechnologist. 'As far as I am aware, there are no genetically engineered Christmas trees,' says Podila. But it can be only a matter of time.

THE CANDLE AT CHRISTMAS

Along with trees and yule logs, another essential component of the Christmas festivities is the candle. No one knows who invented candles but they are the descendants of lamps that burnt liquid fuel, and date back at least 30,000 years to the days when they were used by prehistoric people daubing pictures deep inside caves. Until the risk of deadly blazes became apparent, most Christmas trees were illuminated by the flickering light of candles. Like so many Christmas traditions, the use of candles in this way can be traced to Germany, notably to Martin Luther's decorated tree and to the Advent wreath, a circle of evergreen branches, entwined with red ribbon, holding four candles. When it is bleak outside, the wreath reminds us that there is life and freshness in the middle of winter; its flames also have a religious significance, one candle being lit to mark the passing of each week of Advent. Often a fifth white candle is placed in the centre of the wreath: this is the Christ Candle, symbolizing Christ's birth, and it is lit on Christmas Eve – *Heiligabend* in German – or on Christmas Day.

Light still plays an important role in many Christmas celebrations today. Christians in China use paper lanterns to turn Christmas trees into 'trees of light'. In parts of India, oil-burning clay lamps are used, while in Sweden people offer thanks to the Queen of Light for bringing hope during the darkest time of the year in Lucia processions. Candles are also lit on a seven- or nine-branch candelabrum, called the Menorah or Hanukkiya, during Hanukkah, the Jewish festival of lights held to celebrate the victory of Judas Maccabees over the Syrian tyrant Antiochus more than two millennia ago.

Candles figure prominently in the non-religious African-American winter holiday Kwanzaa, which was introduced in the United States in 1966, inspired by the civil-rights struggles of the sixties. Seven candles, one black, three red and three green, are placed in a candle holder called a Kinara. The candles represent the seven principles on which the festival is based: the black candle symbolizes the African-American people; the red candles their struggle; and the green their hopes for the future. The candles are usually lit one by one before the evening meal on seven consecutive evenings, beginning on the day after Christmas. As each candle is lit, those present recite the associated principle of Kwanzaa, taking turns to give their personal view of its significance.

WHY DOES A CANDLE BURN?

The candle also featured in the most famous of the nineteenth-century scientific lectures for children given over three and a half decades by the pioneer of electricity and magnetism, Michael Faraday. The lectures were held each year during the Christmas holidays at the Royal Institution in London, and culminated in 1860–1 in a series entitled 'The Chemical History of the Candle'. In introducing his lecture, Faraday told his young audience that 'There is no better, there is no more open door by which you can enter into the study of natural philosophy [science].'

Faraday's vision was based on the classical physics of his day, when even the idea of atoms and molecules was not accepted by all scientists. Today, our picture of what goes on in a candle, for instance when it is lit, is informed by the atomic theory of matter: when you hold a match flame near the wick, its heat raises the temperature of the solid wax, which increases the degree of molecular restlessness and agitation. Once the molecules in the wax have sufficient energy to break loose and tumble over each other, the wax melts. We know as much from the change in the appearance of the wax, from opaque to transparent: in the former case, the wax molecules are organized into groups that scatter light, whereas in the latter, the molecules are disorganized and cannot reflect light.

Peter Atkins, the well-known Oxford chemist, points out that Faraday sidestepped some fundamental questions posed by the candle. Why does it have to be ignited with a match? What drives the chemical reaction? This driving force is expressed by the Second Law of Thermodynamics, which can be roughly translated as the tendency of things to become disorganized. In the case of the Christmas candle, matter and energy are dispersed and dissipated in the form of carbon dioxide, water vapour and heat while the candle is burning. The match is required because, despite this natural tendency, there is an energy barrier between reactants in the candle and the products of combustion. Think of this barrier as something like an initial investment that must be made before a candle can enter the energy business. The combustion reaction will only take place if the reactants can be given help vaulting this barrier – by importing some energy with a match to cause reacting chemicals to clash and snap chemical bonds.

The heat of the flame raises the temperature of the candle wax so that its molecules have enough energy to escape into the air. The wax molecules, like all hydrocarbons, consist of a long chain of carbon fringed with hydrogen atoms. These molecules absorb even more energy when they come into contact with heat from the lit match, and vibrate so violently that they shake apart, splintering into smaller molecules, a process called pyrolysis. The fragments react with oxygen in the air. This is fire.

Fuel vapour and oxygen mix at high temperature at the surface of the flame to release heat as new chemical bonds are forged between oxygen and the carbon compounds of the vaporized wax. The resulting rise in temperature melts and vaporizes more wax to nourish the dark heart of the flame. Simultaneously, oxygen from the air moves towards the flame surface by diffusion and convection. An uneasy balance between these two forces shapes the surface of the flame. The process of burning is complicated by the generation of loose atoms of hydrogen, oxygen and their combination, called hydroxyl. These chemical species are called radicals and, as their name suggests, they have the ability to cause chemical mayhem. For example, freed hydrogen atoms bind to oxygen from the air in a reaction that releases great heat. The result: water vapour and carbon-dioxide gas. Both rise from the flame because hot, less dense, products ascend by the process of convection. This imports oxygen and exports combustion products, drawing the flame out into the characteristic shape of a teardrop.

Examine the flame of an Advent candle more closely and you can see that just above and around the burning wick is a dark cone topped by a luminous yellow region. The temperature in the dark cone is relatively low (600°C/1,100°F) and rises (to about 1,200°C/2,200°F) at the heart of the yellow region. The highest temperature (1,400°C/2,550°F) is found on the edge of the yellow portion of the flame. The vaporized wax hydrocarbon molecules are decomposed by the heat in the dark cone near the wick. Around the yellow zone of the flame, radicals from the decomposed hydrocarbons react with oxygen from the air to form carbon dioxide and water in a complex way we do not yet fully understand. A spoon can confirm this. Hold it above the flame and notice how water condenses on the cooler metal; place it in the yellow zone and you can see soot; place it above the wick in the dark zone and vaporized hydrocarbons will condense to form a thin layer of wax. This was one of the many demonstrations Faraday used in his Christmas lectures.

CANDLELIGHT

Given the ancient preoccupation with flames in the cold of winter, it is perhaps not surprising that most of the radiation released by burning is heat, known as infrared radiation. Faraday was at his most poetic when discussing the remaining trace of radiation – light – that is visible to the naked eye: 'You have the glittering beauty of gold and silver, and the still-higher lustre of jewels, like the ruby and diamond; but none of these rival the brilliancy and beauty of flame.'

The luminous yellow zone of the flame responsible for most of the light is called the carbon zone. The soot particles that lurk there are composed of carbon atoms that have been released from their original chains in wax to link themselves into hexagonal arrays of various sizes, from 10 to 200 nanometres, and then into chains. The soot region also contains one of the most fashionable molecules currently under study, a recently discovered form of carbon called buckminsterfullerene, in which sixty carbon atoms link up to form what looks like a football. But how can these carbon footballs and other components of black soot become so incandescent? Soot particles do, indeed, glow like any other hot matter but, as Faraday remarked, 'You would hardly think that all those substances that fly about London, in the form of soots and blacks, are the very beauty and life of the flame.'

At the time of Faraday's lecture, contemporary physics could not account for the yellow and orange colours of a Christmas candle. This would require an understanding of events at the atomic and molecular level. It would take another half a century for primitive models of the atom to emerge, such as the 'plum-pudding' model proposed by J.J. Thomson in 1904, in which an atom was seen (wrongly) as consisting of negative particles embedded like plums within a sphere of positive electricity of uniform density. Today, thanks to quantum theory, the atom is understood to be a tiny positive nucleus surrounded by a mist of negatively charged electrons. The shape and extent of the mist reflect the energy levels of the electrons. When heated, carbon atoms in candle soot shake about, twisting, jumping and then distorting, while electrons hop around between energy levels. Some of this energy converts itself into particles of light which fly out of the flame, varying in colour (frequency) according to the size of the energy hops.

The classical physics practised by Faraday worked reasonably well for low frequencies, at the red end of the spectrum. But at the blue end, it predicted that a heated body would send out an infinite amount of energy. This absurd conjecture was dubbed the 'ultraviolet catastrophe'. The true explanation would have to wait for quantum physics, launched at the turn of the twentieth century by the work of Max Planck at the University of Berlin. To present this supposed catastrophe, and explain how hot bodies such as candles radiate light, Planck had to assume that the atoms in the soot could vibrate only in multiples of a basic quantity, found by multiplying a tiny constant by the frequency of the light.

Then, in 1905, Albert Einstein realized that light could only exist in little chunks, called quanta, a view at odds with the prevailing picture of light as a wave. Two decades later, the American chemist G.N. Lewis christened the quantum of light the photon. The bluer the light, the more energy each photon has. For example, the pale-blue zone at the bottom of a flame is due to quanta sent out by 'excited' versions of water, carbon dioxide and other molecular fragments, which are agitated by the promotion of their component electrons to a higher run on a ladder of molecular energy levels. When these electrons drop down to a more stable level, the discarded energy is sent out in the form of photons.

THE MYSTERY

In one respect, the candle seems almost as mysterious now as it did in the days of Faraday's splendid Christmas lectures – if you are an astronaut, that is. For years, scientists have puzzled over the behaviour of candles in the absence of gravity. Would such a cosmic Advent candle burn indefinitely or be snuffed out by a surplus of combustion products and by oxygen starvation? At the heart of the question is the way in which the lack of gravity affects the flame's supply of fuel and oxygen. On Earth, molecules diffuse and are moved by convection, as the hotter air rises and cooler air descends. In space, however, the process of convection is switched off. The flame should burn, but only so long as the oxygen molecules diffuse inwards quickly enough and the hot fuel diffuses outwards.

Scientists are divided as to whether the flame of a Christmas candle could survive for very long in zero gravity. Some argue that it could, according to the equations governing the process of diffusion. Others believe that diffusion is too slow a process to nourish the flame. Experiments on Earth, which involved literally dropping the candle to switch off the effects of gravity, have proved inconclusive.

Another method to turn off gravity was adopted by Howard

Ross and Dan Dietrich at NASA's Lewis Research Centre in Cleveland, Ohio, working with James T'ien of Case Western Reserve University in Cleveland, Ohio. They studied a candle flame aboard a 1992 shuttle mission. Immediately after being lit, the flame was spherical and had a bright-yellow core, presumably from soot. After 8–10 seconds, the yellow disappeared and the flame became blue and measured only half an inch or so across. There being no gravity to define an 'up' or a 'down', a space flame tends towards sphericity. However, the shuttle experiment showed that heat lost to the wick of the candle quenches the base of the flame, causing it to be hemispherical. In the shuttle, the process of diffusion controls the supply of oxygen and fuel vapour to the basin-shaped flame. Being a slower process than convection, this reduced the heat generation so the temperature is lowered to the point where little or no soot forms, as evidenced by the all-blue flame.

Further investigations were carried out aboard the Russian space station *Mir* in 1966 by American astronaut Shannon Lucid. So far, the studies seem to suggest that a candle can burn in the 'microgravity' of space; however, the quest has raised other questions that have still to be answered, such as the reason for the presence of strange flickerings just before a space candle is extinguished. Despite scientific investigations, the flame retains some of its mystery and inspirational quality. The latter was expressed by Faraday, who, echoing the prayers familiar from Christmas church services, hoped that we might all be compared to a candle: 'That you may, like it, shine as lights to those about you; that, in all your actions, you may justify the beauty of the taper by making your deeds honourable and effectual in the discharge of your duty to your fellow men.'

Six

Giving and Shopping

> The only gift is a portion of thyself. Thou must bleed for me. Therefore the poet brings his poem; the shepherd, his lamb; the farmer, corn; the miner, a gem; the sailor, coral and shells; the painter, his picture; the girl, a handkerchief of her own sewing … It is a cold, lifeless business when you go to the shops to buy me something, which does not represent your life and talent.
>
> Ralph Waldo Emerson, 'Gifts' (*Essays*)

Even the apparently innocent act of giving a Christmas card groans with meaning. According to psychologists, this ritual reveals a great deal about our nature and our relationship with others. The same goes for giving presents – and even the way we wrap them. We moan endlessly about the 'materialism' of the Christmas holiday, the consumer hype, the hard-sell advertising. Yet Christmas is commercially more important than ever. Christmas shopping now starts in late summer and, by one estimate, purchases account for one-sixth of all retail sales in the United States

In Britain around 8 per cent of the economy is devoted to producing articles that will be given as presents, with the cost of Christmas gifts amounting to 4 per cent of an individual's annual income. In Japan, where shoppers scurry around during the *oseibo* gift-giving season to the refrain of *'Meri Kurisumasu'*, the proportion is probably greater. Meanwhile, anthropologists tackle the paradox of the seasonal spending spree. How can Christmas be both a religious festival and the greatest commercial holiday in the Christian world? This paradox is no accident, argue the anthropologists. Shopping and giving are part of the Christmas festivities because they help to reaffirm family values in an increasingly materialistic and impersonal world.

The American organization SCROOGE (Society to Curtail Ridiculous, Outrageous, and Ostentatious Gift Exchanges) has called for people to purchase self-improvement lessons, smoke alarms, first-aid kits and other *sensible* gifts. But, as one academic pointed out, this cold-eyed, rational view of Christmas could do more to demolish the romance of the event than the annual ritual of mass consumption itself.

Of all the Christmas stories written by Charles Dickens, *A Christmas Carol* (1843) conveys most vividly not only the Victorian Christmas but the development of the modern festivities, with their spirit of lavish spending. The story of Ebenezer Scrooge is the best-known secular Christmas tale on both sides of the Atlantic. Scrooge's transformation from a heartless capitalist who regards Christmas as a frivolous interruption of his money-making mission, into a spendthrift who tips the newsboy and lavishes gifts on the poor Cratchit family, encapsulates the most important component of modern attitudes to Christmas. Scrooge's diversion from the path of avarice to that of altruism symbolizes a turning-away from the thoughts of self towards thoughts of others. This theme of charity, community and the long-term benefits of kindness has proved inspirational ever since the book was published.

When Dickens read the *Carol* aloud in Vermont in 1867, it made a deep impression on one of those assembled before him. A Mrs Fairbanks noted that her manufacturer-husband's face 'bore an expression of unusual seriousness'. Afterwards, perhaps struck by his resemblance to Scrooge, he declared that he intended to 'break the custom we have hitherto observed of opening the works on Christmas Day.' From then on, each Christmas Eve he sent his factory workers home with a turkey.

Robert Louis Stevenson declared of Dickens's Christmas

books: 'I have cried my eyes out, and had a terrible fight not to
sob. But oh, dear God, they are *good* – and I feel so good after
them – I shall do good and lose no time – I want to go out and
comfort someone – I *shall* give money. Oh, what a jolly thing it
is for a man to have written books like these and just filled
people's hearts with pity.' Today, basking in the afterglow of *A
Christmas Carol*, Christmas past still seems brighter, warmer,
happier and, well, 'Christmassy' compared with the modern
celebration. This may simply be because events remembered
from our childhood are imbued with nostalgia. However,
many people also grumble about the way in which the modern
Christmas has become commercialized. This complaint has
been made many times before – indeed, so often that it is very
much part of the festivities.

Eighty years ago, a London *Times* leader looked back nostal-
gically to the typical Dickens Christmas, with its high spirits,
charity and simple pleasures. But Dickens points out in *Sketches
by Boz* that some will 'tell you that Christmas is not to them
what it used to be.' The author of *The History of the Christmas
Festival*, published in 1843, also wrote of bygone times when
there was a greater emphasis on old customs and hospitality. A
few years before that, Sir Walter Scott recalled when 'England
was merry England'. Go back still further to 1790 and you find
a leader article in *The Times* reporting that 'Within the last half
century this annual time of festivity has lost much of its original
mirth and hospitality.'

The Times did not say much about Christmas again for the
next fifty years or so. It seems that the Victorians showed a
renewed interest in the season, spurred on by a nostalgia for ye
olde yuletide traditions of Christmas past. From the mid-1830s
onwards, this resurgence of enthusiasm was reflected in the
increasing coverage given to the event by newspapers and peri-
odicals. The 'traditional' Christmas was epitomized by Charles
Dickens's descriptions in *The Pickwick Papers* (1836) and *A
Christmas Carol*, and by Washington Irving in his *Sketch Book*

(1818), in which he bemoaned how 'many of the games and ceremonials of Christmas have entirely disappeared.' One of the sad ironies of Dickens's career is that he who so loathed the hollow nostalgia of 'the good old days' is now remembered not only as a great novelist and social critic, but also as the creator of the picture-postcard Victorian Christmas, to which we look back with dewy-eyed regret.

Even the greeting 'Merry Christmas!' seems to carry an echo of 'merrie England' and a world where the snow lay deep and crisp and even, the home was decked out with holly and ivy, and a magnificent tree glittered in the front window. No wonder Christmases don't always live up to our expectations.

A few universals lie at the heart of Christmas, however, no matter where it is celebrated: the birth of Christ, the home, family, children, and, of course, an orgy of giving. Indeed, you may have been given this book as a Christmas present! I certainly hope so. Russell Belk of the School of Business at the University of Utah notes that when compared to birthday gifts, Christmas presents are more practical, more prestigious, more personal, more expensive, more fashionable, of higher quality, and longer lasting.

The language of giving is ancient. In the animal kingdom, gifts are made for good reason. In many insect and some bird species, females accept males as mates only if they are offered a morsel of food. Collecting such a gift may be as hazardous for males as the hunt for a female itself. For example, male scorpion flies mate only if they offer females 'nuptial gifts' of palatable insect prey. In searching for these presents, the males risk becoming entangled in the webs of predatory spiders. Such generosity may seem baffling when viewed in the light of Darwin's theory of natural selection, which seeks to explain the evolution of all sorts of adaptations that increase an animal's chances of survival against the backdrop of 'nature red in tooth and claw'. Darwin was himself puzzled by behaviours that do nothing to promote survival and may even threaten it. He concluded that features such as the peacock's tail – characteristics usually sported only by males – evolve by a process called sexual selection. Individuals possessing these desirable characteristics acquire more mates than their competitors of the same sex and thus have more chance of reproducing. For those 'generous' birds, insects and other creatures, gifts therefore have a well-defined purpose: to help them to pass on their genes.

Human beings also give presents for a purpose and have been doing so since the emergence of hunting technology some 50,000 years ago. Darts and other projectile weapons enabled hunters to tackle game whose size made it necessary to share the spoils (a mammoth, for instance). Our generosity is only partly due to a desire to be nice to our fellow man and woman. We expect something in return.

Giving at Christmas is charged with meaning. In 1252, Henry III entertained a thousand knights and peers at York; the Christmas feast was so costly that the local archbishop donated 600 oxen and £2,700 – a vast sum for the time – towards the feasting. A century and a half later, Richard II provided 2,000 oxen and 200 tuns of wine for 10,000 guests at Christmas. Edward IV fed more than 2,000 people each day at Eltham over the Christmas period of 1482. This lavish hospitality was extended for a good reason: to bind alliances and strengthen community bonds. In this way, kings underwrote feudal relationships. Part of the seasonal package was for the powerful to attend to the needs of the powerless, for instance by serving them dinner, an inversion that helped to ease tensions between the strong and the weak and thereby reinforce the status quo.

Gift-giving in pre-industrialized societies was seen by the French ethnographer Marcel Mauss as a way of forging social contracts with strangers. In his *Essai sur le don* (1925), regarded by anthropologists as a masterpiece, Mauss explained how generosity served a useful purpose in the days when there was no state police or militia to keep the peace. He reminds us that there is no gift without bond or obligation.

In his book *The Origins of Virtue*, the science writer Matt Ridley brings the ideas of gift exchange and reciprocity to life with a description of the Hazda people of Tanzania. In the wooded savannah near Lake Eyasi they forage for honey, roots and berries and hunt large animals like giraffe. When it comes to the big game they either return empty-handed, which is usually the case, or they down an antelope or giraffe with a bow and arrow, only to give most of the meat away. What the hunters get in return for their generosity is kudos and the kind of female admiration that could pave the way to extramarital affairs, underlining the fact that anthropologists don't quite equate tit-for-tat with reciprocity: the former means swapping similar favours at different times, the latter sharing different favours at the same time.

The Hazda hunters are converting giraffe meat into a durable and valuable commodity – prestige – that will be cashed in for a different currency of advantage at a later stage. Ridley sees in these actions echoes of the origins of the financial derivatives markets: '[The hunter] is entering into a contract to swap the variable return rate on his hunting effort for a more nearly fixed return rate achieved by his whole group.' This kind of thinking transforms the meaning of the gift from one of social sentimentality to one of impersonal economics that follow culturally prescribed laws and obligations. Indeed, in some societies the sense of obligation attached to the reciprocal gift was so strong that the gift could be used as a weapon.

In the Pacific North-west, until the nineteenth century, the indigenous 'first-nation' people would attempt to belittle their rivals by the so-called potlatch ritual. In a status-obsessed society, rivals staged a bizarre battle of generosity in which blankets, oils, food, pelts, canoes and decorated sheets of copper were given away: every present meant another rung up the social ladder and every failure to reciprocate meant loss of status in these escalating gift wars. Many commentators have remarked that this competitive gift-giving sounds rather like the exchange of presents that goes on every Christmas in our own society.

In the sixties the anthropologist Marshall Sahlins noted that the closer the kinship between the donor and the recipient of a gift, the less emphasis was placed on reciprocity and the more on sentiment. Within the family, there was little tally kept of who spent what on whom. Within the larger social grouping of the tribe, however, a certain amount of diligence was required to ensure that you gave as good as you got. This same reciprocity ruled when it came to gifts between unrelated allies. But in the case of rival tribes, the idea was to try to get more out of them than they got out of you.

Echoes of what Sahlins had noted were observed in Middletown, a community in the Midwest of the United States much studied by ethnologists. There, in the early eighties, Christmas gift-giving was studied by Theodore Caplow of the University of Virginia. He found that the vast majority of presents went to those within the nuclear family. Gifts to relatives outside the core family were less common and more conditional. For extended family – siblings, siblings' spouses, siblings' children, and parents' siblings – giving was less frequent still. Gifts within the core family were given without the expectation of an equivalent return, but Caplow found a growing emphasis on reciprocation when it came to the extended family. And beyond the extended family the Middletown study revealed mostly small gifts, amounting to little more than tips for tradesmen or teachers. Another study by Caplow, based on interviews with 110 people in Muncie, Indiana, found that when people did not give to someone who had previously been on their list of recipients, they considered that the relationship had waned in importance, and might even end altogether. The non-appearance of a gift can place a firm full stop at the end of a friendship.

THE PSYCHOLOGY OF THE GIFT

The giving of gifts becomes a dialogue of intimacy, a language for the communication of emotions and values. Gifts have the power to make or break a relationship; they can infer status, power, taste and emotion. And, according to Adrian Furnham, a psychologist at University College London, they reveal how socially aware we are in our perceptions of others. It is not just the problem of who to give presents to, or how much or how little you spend on those presents, but of what sort of gift you select. And when our motives for giving a particular gift are incorrectly interpreted, our faux pas is on show for all to see. For instance, the gift of a fluorescent fluffy toy might be thought an insult by someone who perceives themselves as sophisticated. 'As a channel of communication [the gift] has limited capacity because the range of messages is few and the language not well known,' says Furnham. 'Perhaps the gift-phobics who discourage the exchange of gifts between family and friends do so because they don't speak the language and agree with Wittgenstein, who so wisely noted: "Whereof one cannot speak, thereof one must be silent."'

Men and women reveal different attitudes towards this seasonal ritual. At the University of Winnipeg in Manitoba, the sociologist David Cheal had great difficulty finding as many men as women to interview for one study of Christmas gift-giving. The reason soon became clear: despite sex equality, the annual hunt for that ideal present is overwhelmingly seen as 'women's work'. Indeed, among couples it is usually the women who maintain the 'gift economy'. Men, on the other hand, tend to give more valuable gifts, less often. Part of the reason is that personal income is usually higher for men than for women, but this is not the whole picture. Women may perhaps be said to dominate Christmas because it is seen as a family festival and women are the 'kin keepers' who take most responsibility for maintaining family and social ties. One of the women in David Cheal's surveys explained that her reason for giving is 'to be a message. You have interest in that person. You have concern. You have love, affection, whatever the message is at the moment.'

Other studies have shown that, in many respects, we are little different from the Hazda of Tanzania. Gift-giving will often put somebody else under an obligation, exploiting a reciprocal instinct that brings the act closer to pure barter. Christmas giving follows certain rules and obeys certain taboos, say Carole Burgoyne and Stephen Lea of the University of Exeter, England. One traditional taboo is the gift of money, which implies a lack of effort and insight on the part of the giver, according to a research project conducted by Burgoyne, with David Routh of Bristol University, involving ninety-two students. Another study by Stephen Lea showed that money is considered particularly inappropriate when given from child to parent, though perfectly acceptable when it is a gift from a grandparent or parent to a child.

Today's psychologists see gifts as a way of initiating and maintaining relationships – just as in the case of Henry III. According to Carole Burgoyne, Christmas tends to differ from other giving rituals, such as birthdays, because gifts are more likely to involve a simultaneous exchange. In relationships where reciprocity is expected there can be serious consequences of a failure to give a gift. 'the non-appearance of a gift is likely to lead to broken relationships and family rows unless there is a very good explanation for it,' Burgoyne says.

Nevertheless, Christmas sees the relaxation of other rules of gift-giving. Because they are handed out more widely, seasonal presents are often less intimate and personal than birthday presents. This, of course, can be an advantage for those who are trying to start up a relationship. But gently does it, warns Burgoyne: 'Gifts that are too expensive may signal a level of commitment and impose a sense of obligation that is not wanted by the recipient. Thus an inappropriate gift – one that is either too cheap or expensive, or one that seems to expose a lack of taste on the part of the donor – carries the risk of rejection.'

THE PSYCHOLOGY OF THE CHRISTMAS CARD

The most common Christmas gift of all is the Christmas card. Henry Cole's nineteenth-century prototype paved the way for a vast variety of cards, some dressed in satin and silk, others gilded and frosted, or fashioned in the form of stars, crescents and fans. In the United States, the very first Christmas card appeared some time between 1850 and 1852 as an advertisement for 'Pease's Great Varety (*sic*) Store in the Temple of Fancy', adorned with Santa and a black slave laying the Christmas dinner table. In the 1880s, manufacturers began producing cards cheaply, enabling anyone to send them. Cards were also sold in packs, one example being the 'Penny Basket' introduced by the German company S. Hildesheimer & Co. By 1883, *The Times* noted that this 'wholesome custom' had become a way to end strife, mend relationships and strengthen family ties.

Today's psychologists would agree, only more so. Cary Cooper, an American professor of psychology working at the University of Manchester Institute of Science and Technology, notes that the act of sending a card is laden with meaning. Even as late as the seventies, exchanging Christmas cards was little more than a minor social obligation. Today, after the 'I-driven' culture of the eighties, it has become far more significant. 'We are entering an era where people are thinking more about other people,' says Cooper. When we send cards, we are reaching out to those with whom we have lost contact, or with whom we want to sustain a relationship. 'It is a form of social networking,' he adds. In an age when many people no longer live near their extended families, do not know their neighbours, and are highly mobile, cards provide social glue. 'The cards I get from the United States don't just have signatures but contain letters and pictures,' Cooper notes. 'It is a way of saying, "Hey, I haven't forgotten you. I still value your friendship."'

The cards also provide an insight into the Christmas psyche. Just like the decisions we make about whom to vote for, whom to marry, and what to wear (things we think of as purely rational, individual matters), the Christmas cards we send and the way in which we send them may, in fact, say a lot more about us than we would care to admit. Ironically, this may be another reason why the custom has become such a central feature of Christmas. In a recent analysis, Adrian Furnham and Ruth Leigh of University College London point out that there are many things going through our minds when we devise a list of whom to send cards to. A few people will have kept last year's crop of cards so they can religiously repay their debts. Some people scour their address books for all those with whom they have lost touch. Others fit the chore in among the obligations to shop or put up decorations, or simply send cards tit-for-tat as other cards drop through their letterbox.

Any deviations from a straightforward exchange reveal a lot about status, argue Furnham and Leigh. People to whom you send cards while receiving none in return are likely to be above you on the social ladder. Those from whom you received cards but to whom you could not find time – or be bothered – to send cards back are probably below you. They are 'likely to be attempting to ingratiate themselves with, or gain favours from, you yourself,' note Furnham and Leigh. 'Of course this is more exaggerated in individuals who are upwardly socially mobile.' What is written on the card is also revealing. A long, rambling message is often penned by someone trying to impress. A little joke may attempt to make up for the long delay in re-establishing contact. A simple signature is saying that the sender is much too busy and important to bother with pleasantries.

The type of card chosen also betrays a great deal of

information about the person who sent it. A home-made card may say: 'I am wealthy and have enough leisure time to make my own cards.' Or it may declare: 'Admire my artistic bent!' Or: 'I am spending all my time on your card to show you just how much I care for you.' A 'green' card printed on recycled paper and carrying a greeting in a foreign language can not only establish the sender's environmental credentials but also suggest that the recipient should feel a twinge of guilt for sending that commercial card on non-recycled paper adorned with reindeer, Santa, or a mouse peeking out of a stocking. A privately printed card reeks of money or self-importance – the sender is too busy to sign a lot of cards. An interesting variant is the photograph card, which is calculated to show off that new house or new addition to the family, whether car or baby. Then there is the institutional card. As Furnham and Leigh remark, 'This has much of the same function as people who wear allegiance ties, cuff links, blazer badges or some other form of insignia, stating publicly their identification with the values and aims of an institution.'

There are many other types of Christmas card: the political card, used as a badge of affiliation; the cartoon card, for those who have no desire to remember what Christmas is all about; and cards with Dickensian scenes, mythical figures, Europeanized Biblical scenes and anthropomorphized animals. There is even the microchip musical card: 'Their tinny warblings can only be stopped by violent means, and to make things worse, the picture will either be of a fluffy animal or a badly drawn snow scene. Sever all relationships with people who send these cards,' warn the University College psychologists. Finally, of course, the spirit and influence of Dickens is alive and well in the form of the charity card. This advertises the generosity of the sender, suggesting even – heaven forbid – that they are suffused with the true spirit of Christmas, brotherhood, peace and generosity. Because these cards enable one to do so many things simultaneously, they are becoming the most popular.

How you display your cards is also revealing. The fact that they are on show at all is a way of announcing to visitors how many friends and acquaintances you have. A century ago, the calling card of any rich or famous person you had encountered would have been left in the hall for visitors to marvel at. Today, the same favour is bestowed on celebrity cards, which are prominently displayed on the mantelpiece alongside those from anyone who may happen to pop in to offer Christmas greetings in person.

AN ANTHROPOLOGICAL CHRISTMAS

Anthropologists believe that *they* are the ones who really understand Christmas. In a milestone study, *Unwrapping Christmas*, edited by Professor Daniel Miller of University College London, they point out that the emphasis that Europeans and Americans now place on emotionally charged gift-giving emerged in the mid- to late nineteenth century, coinciding with intensive industrialization and a growing sense of the domestic sphere as a special moral realm distinct from the harsh realities of the workplace. This emphasis on the 'family' Christmas was spurred on by the enormous popularity of Dickens's *A Christmas Carol*.

However, the custom of buying manufactured goods as Christmas gifts did not become widespread until the 1870s. In the United States, leading department stores such as Macy's in New York began to lure customers with striking displays of imported dolls and other exotic consumables. Santa Claus also began to make his appearance in those stores. 'If Santa is the god of materialism, what better place to enthrone him than in the department store, which did so much to foster consumer culture,' comments Russell Belk of the School of Business at the University of Utah.

We are now living in a 'commodity' culture, one in which we use money to buy impersonal, mass-produced objects made for profit by people we've never met. As the anthropologist James Carrier of the University of Virginia puts it, these objects 'bear no human identity or being. They are manufactured in the world of work and express only the impersonal desire for profit by the company that made them and the impersonal acting out of work roles, lightly adopted and in return for money, by the unknown people who produced them.'

The challenge we face at Christmas, argue the anthropologists, is to transform the mass-produced objects found in stores into gifts by scrubbing them clean of the contamination caused by their association with money. For example, we may mutter 'It's the thought that counts', and place the emphasis on gifts that are frivolous rather than useful. So, the argument goes, it is better to purchase luxury items than 'necessities' that might be bought at other times of the year – a bottle of expensive perfume rather than a frying pan, for example. Another approach is to remove the price tags, wrap the gift in patterned paper and tie it up with ribbons and bows. Some people joke about spending more time and money on the wrapper than the contents, and this is often the case when gifts are given in Japan. By doing this, the giver temporarily makes the object irrelevant and the act of giving the focus of attention. The anthropologist Claude Lévi-Strauss argued that wrapping overlays the mass-produced product with sentiment and with the giver's identity.

Anthropologists mutter about 'decontaminating' products that have been bought in the marketplace, a worldview that carries the shocking implication that we had to invent the ritual hell of Christmas shopping as part of this cleansing process. James Carrier argues that shopping is integral to the experience rather than 'an unfortunate commercial accretion on a real ritual and familial core'. In this respect it is like the ritual of family cooking, which converts store-bought groceries into a meal that expresses, embodies and strengthens the family bond. The next time you whine about the crowds of shoppers in December, just remember that the seasonal spending spree is *meant* to be unpleasant. Christmas has to be 'worked at' if it is to be done 'properly', says Carrier, and the Christmas-shopping scrum is required as a sacrifice of time and effort to ensure that gifts have meaning. That is why people grumble endlessly about what hard work the shopping is, and about the growing commercialization of Christmas. These complaints help to affirm that at least once a year it is possible to 'wrest family values from recalcitrant raw materials'.

Theodore Caplow's study of Middletown in the United States revealed that few Christmas gifts were handmade, and most of them were given, unsurprisingly, by young children. In other countries the situation is different. In Sweden, home-made gifts, including home-pickled cowberry sprigs, are *de rigueur*, precisely because people are anxious about 'folk culture' and have a strong sense of national identity. In North America and Britain, by contrast, home-made gifts are eschewed because in these countries anxiety centres on how to make 'family values' compatible with rampant commercialism.

Paradoxically, Christmas can even be regarded as a 'festival of anti-materialism', argues Daniel Miller. The obsession with personalizing gifts counteracts what we think of as the all-too-intimate relationship between Christmas and materialism. 'We glut, we overdo it, we grumble about how commercial it all is, but we are reflecting back on the value of it all,' says Marilyn Strathern, Professor of Social Anthropology at the University of Cambridge.

BIG BROTHER IS WATCHING YOU SHOP

Anthropologists are not the only ones who watch your every move as you hunt for that ideal gift. They have been joined by psychologists armed with computers, scanners and video cameras, whose purpose is to help stores extract even more money from you during the seasonal spend. Computers and cameras study customers' behaviour, while retailers experiment with smells, music and 'atmospherics' to encourage people to buy just one more Christmas gift.

However, the most obvious incentive for shoppers is the offer of lower prices in a pre-Christmas sale. Robert East of Kingston University in London notes that a drop in price of 10 per cent can boost sales by 20–30 per cent. If the goods are also placed in a special display, sales can go up by 80 per cent. Add local advertising, and figures can be pushed to 200 per cent in some circumstances. There are various strategies used by retailers in the markdown of Christmas merchandise, adds Dale Lewison of the University of Akron, Ohio, and timing is crucial: 'The timing of markdowns becomes quite an art and science if the retailer wants to maximize both revenues and profit.' The longer the retailer waits, the bigger the price reduction needed to generate customer interest. Many retailers have a policy of marking down goods every week or two weeks.

Stores can also increase sales by installing elaborate window displays with animated Santas, electronic elves and mountains of fake snow. To find out how effective such displays are, the Retail Analysis Team at Nottingham University in England used video equipment to film shoppers. 'The idea was to see how far away people walking along the pavement are attracted to displays and whether they are tempted to enter,' says Roy Bradshaw, a member of the team. More than a million observations were taken of pedestrians in city shopping areas and in stores. This huge study showed that as many as five times more people entered some stores than was recorded by till receipts. It also suggested that a thirty-year-old mother will behave differently if she is shopping midweek in her lunch break for essential items than if she shops at the weekend with her partner and/or young children.

The Nottingham University data revealed where shoppers look, where they go, the bottlenecks and the number of sales per thousand customers passing the product. Many stores have a wide path leading from the entrance that is supposed to draw punters in a U-shaped track around merchandise displayed on 'gondolas', or island shelves. Some seasonal shoppers tended to 'bounce' around the track or sometimes pass through a store without buying. For retailers, this finding was somewhat shocking. The studies have also shown that in some cases an improvement in sales of as much as 20 per cent could be made by altering store layout.

Another shopping spy comes in the form of the bar code, which has been monitoring shoppers since 1967, when the first retail system was introduced into a Kroger supermarket in Cincinnati in the United States. Six years later the retail industry agreed on the Universal Product Code, which today is to shopping what DNA is to biology. The pattern of lines and bars in the code can reveal to a laser scanner the identity of a product. Thin lines at either end of the code denote the manufacturer and item; a database is then accessed to pull out the price, check how many are in stock and so on.

Using a computer, a bar code can be combined with information about the layout of the store to highlight hotspots of activity. Unsurprisingly, such analysis shows that eye-level height is the most profitable for a display of goods, giving about twice the return of facings at a lower level. Demand can also be linked with weather conditions, so that it is possible to show to what degree a cold spell can influence the buying of warm clothing.

Sunil Gupta, who conducts bar-code research at Columbia University, has shown how regular seasonal sales undermine brands in the long run, as customers refuse to buy them at the normal price. These are just a few of the uses to which bar-code data have been put. Eventually, sophisticated computer models will be used by major stores to predict what items to stock and promote each Christmas, drawing on a vast base of information ranging from the weather forecast to customer demographics. Gupta says that department stores are only just beginning to recognize the power of bar-code data.

BIG SHRINK IS WATCHING YOU

The way stores are designed, gift ideas presented and so on is based on a thorough understanding of consumer psychology. This makes the average shopper sound like a hapless pawn ruthlessly manipulated by multi-million-dollar businesses. But, says Mark Uncles of the University of New South Wales in Australia, 'This is not always as sinister as it might appear.' While designers and retailers attempt to influence the moods and emotions of shoppers, sound operational considerations and the needs of the customer are also taken into account. For example, it is more convenient to locate in-store bakeries and fishmongers against the back wall – it isn't simply that these added-value services draw shoppers through the store and stimulate impulse buying.

Nonetheless, there are many intriguing initiatives to help shoppers part with their money by the manipulation of 'atmospherics', from the colours, smells and lighting in the store to its layout, the positioning of gondolas, posters and in-store 'theatre', and the playing of music. For example, supermarket layout often steers shoppers clockwise: 'Psychologists say we have a predisposition to turn right on entering closed spaces,' says Mark Uncles. Jewellery, cosmetics and perfume are displayed on a department store's ground floor to look attractive and create a pleasant smell. The use of the aroma of coffee and freshly baked bread to lure customers goes back to at least the eighteenth century, while the use of fresh herbs is even older, possibly dating from the Middle Ages. Today, many large stores on both sides of the Atlantic pump the smell of fresh bread from the ovens to the entrance of the store to attract customers and set stomachs rumbling. Other stores use the fragrance of freshly ground coffee to pull in those credit cards and cheque books.

In the weeks running up to Christmas, the scent of cloves, cinnamon, brandy, Christmas pudding and other 'jingle smells' may also waft through the aisles. For Christmas 1995, Woolworth stores on both sides of the Atlantic used a system developed by BOC Gases to release a puff of mulled wine dissolved in carbon dioxide, each quarter of an hour. A panel of staff had decided that mulled wine created the right ambience to put customers in a suitably festive mood. Meanwhile the Tate Gallery in London plumped for the smell of brandy.

Although it has been used for hundreds of years, smell has been remarketed as the 'hidden persuader' in the light of research that suggests that the right smells might encourage consumers to spend more money. One study of sales at a women's-wear retailer in the United States found that there was a 15–20 per cent increase in the average amount customers spent when the store was filled with a peach aroma, compared to when no scent was used. Earnings in a Las Vegas casino jumped by 50 per cent when a pleasant smell was wafted in front of gamblers, and shoppers were found to be more willing to buy Nike sneakers, and even to pay more for them, in a floral-scented room.

Smell is not the only sense that is being manipulated by retailers. Colours are also thought to help part you from your money. It is widely believed that people find red more arousing than 'cool' colours such as blue and green. This has led to the suggestion that red is appropriate where quick decisions will benefit the store. Could that be why it is Santa's favourite colour?

To focus attention on Christmas products, stores play seasonal music, especially carols. The pace of the music can be important. Low-tempo 'easy listening' relieves stress and encourages punters to take more time. One study found that slower music boosted sales by 38 per cent compared with those made when faster music was playing. The type of gift purchased can also be influenced by music, according to a study conducted in a supermarket by Adrian North, David Hargreaves and Jennifer McKendrick from the Music Research Group at the

University of Leicester in England. French wine flew from the shelves on days when French accordion music wheezed from the in-store sound system: five times as much French as German wine was sold. But German wine triumphed on days when thigh-slapping *Bierkeller* oompah-pah music poured forth, with twice as much German as French wine being sold, despite an overall sales bias in favour of French wines. This study supports what psychologists call the 'preference-for-prototypes' model. 'If you hear music that you label as French you start to think of things about France in general,' says Adrian North. Other studies have underlined the influential role of music. Classical music prompted the purchase of more upmarket wines than pop music, for example, and a crucial rule for stores where younger people tend to shop is to turn the volume up. But there has to be a limit. The Leicester team is now studying the long-term effects of Christmas carols, notably the irritation caused when shoppers are subjected to them again and again and again.

Atmospherics can also be used to 'demarket' a product, even at a time like Christmas. The classic example is the state-run liquor store, which is deliberately made uninviting in some countries to discourage drinking. You might think that Christmas crowds would have a similar effect, creating log jams of customers, fighting in the aisles, or even 'trolley rage'. Robert East of Kingston University points out that the impact of store scrums is a subtle one: crowds are undoubtedly worst during the Christmas shopping period, but that is not necessarily a bad thing. Whether we like crowds or not depends on where we are: yes, if it is a bar; no, if it is a bank, says East. When it comes to a store, too few people may advertise the fact that it has poor products and high prices, while crowds could make us more purposeful and efficient in our shopping. Equally, they could make stress hormones soar to unbearable levels.

THE LAWS OF CHRISTMAS SHOPPING

An ambitious project to tease out the details of how we hunt for that special Christmas present is being conducted at the Department of Marketing at Concordia University, Montreal. There, researchers have put forward eight propositions that govern whom we consult about buying a gift and how long we spend searching for it. The researchers were interested in studying the effects of a range of factors such as 'experiential pleasure' (the sheer enjoyment of browsing during the gift quest), the shopper's 'social risk reduction strategy' (for instance, how much effort we invest in selecting a gift to avoid appearing cheap or insensitive) and 'psychological risk' (the strength of our resolve to cope with the misery of fighting with a scrum of shoppers). Drawing on earlier research, and a great deal of common sense, they put forward the following 'laws of Christmas shopping':

1 The shopper will spend less time hunting if he/she is accompanied but more time when searching for a costly gift or when a tight budget has been set.

2 The less time we have to shop and the more money we have to spend, the more we rely on information desks and shop assistants.

3 The use of predetermined gift selection – orders from a 'significant other' for a particular brand of perfume, for example – 'result in less in-store general information search, and more brand or item-related information search'.

4 The closer your relationship with the recipient of a gift, the more likely you are to know what they want and the less time you will spend rooting around.

5 The more difficult the recipient, for instance a much-disliked in-law, the more time will be spent finding the gift.

6 Those who love shopping, bargain hunting and giving will

invest more time in the ritual, while those who prefer buying generic items to specific brands will invest less.

7 Variables such as sex, age, education and family size will also influence the hunt for presents. The Concordia researchers found much evidence to support the common observation that male Christmas shoppers are a sorry lot: 'Men will tend to leave their Christmas shopping until the last minute, and then will enter a store, list in hand, desperate,' they comment. One earlier study noted that men were distinguished by 'their desperation, tardiness, and discomfort with the ritual process'. And a recent study conducted at a London shopping mall revealed that male shoppers showed a more consistent surge in heart rate and blood pressure than women, suggesting they were more stressed.

8 The reliance of shoppers on displays – one example of what the team refers to as a 'non-personal in-store information source' – is influenced more by the location of the display, for example, than by demographics such as the social class and potential spending power of the shopper.

Once the Concordia team had drawn up these propositions, they put them to the test. Socks, shirts, scarves and so on had been identified as popular gift choices by an earlier study on Christmas shopping, so the team focused on the quest to buy an item of clothing. To find out what was going through the minds of the shoppers during this quest, a list of questions was prepared, ranging from the religious persuasion of the shopper to the identity of the gift recipient. Of more than a thousand questionnaires distributed, 366 were returned, enough for the team to carry out statistical tests to tease out the relationships between the various factors swaying gift choices.

Their first draft of the laws of Christmas shopping received broad support. For example, women did indeed put more effort into rooting around for the ideal gift than men, and more effort was expended by the parents of older children living at home. Unsurprisingly, people were found to spend more time and were more willing to consult an assistant when seeking a costly gift, such as a mink stole.

There were some interesting exceptions, however. The survey undermined the suggestion that a companion helps. They don't, particularly if the companion is a husband who has been dragged along on the expedition. As the team concludes: 'Women often shop with their husbands, and ask for their opinions, but in truth these wives only want confirmation of their decisions, and do not fully trust their husband's opinions.' Nor do gift lists seem to make much difference when it comes to the amount of time invested in shopping: even if you are told to buy a jumper for a present, that still leaves plenty of designs and styles to choose from. And the shopping rule that people would invest more time in 'psychologically difficult' gift recipients, such as the dreaded mother-in-law, was not backed up by the survey.

Other associations emerged, some unexpected. Consumers with more intense religious beliefs tend to rely more on information provided by the store, suggesting that they have a greater desire to come up with the right gift than infidels. Those with smaller families tend to seek more help from assistants. Bargain hunters will scour the store for information but avoid salespeople, presumably because they don't trust their advice.

This kind of insight into the Christmas-shopping experience can be exploited during the annual spending spree. For example, if stores want to attract men, they should spend relatively little money on packing, signs or merchandise displays. Instead, assistants should be out in force to help male customers make their minds up. For women, however, the converse is true: they need better labelling and fewer intrusive assistants. The team at Concordia are the first to admit that we still have some way to go before we can uncover all the motivational forces involved in Christmas shopping, but they believe they have taken the first step towards defining them.

QUEUES AND SOCKS!

One depressingly familiar Christmas experience is being stuck in a supermarket queue, burdened with vast quantities of food and drink. Fortunately for the seasonal shopper, we now know more about this misery than ever before, thanks to nine decades' work by mathematicians studying the behaviour of queues. Although supermarket queues are all subject to random delays, on average they will tend to move at about the same rate. This feature of the Christmas-shopping experience is captured in a mathematical form by the so-called Poisson process, which assumes that people are as likely to arrive at one time as any other, but that precisely when they arrive is entirely random.

This theory has some subtle implications. For example, you might think that to keep supermarket queues down to a minimum, the rate at which checkouts deal with customers should match the rate at which they arrive. 'Wrong!' says Robert Matthews of Aston University in England. Unless the checkouts process customers faster than they arrive, the queue will just grow and grow over time – in theory, becoming infinite. The reason, Matthews says, is that the checkouts must be able to cope with the randomness in the arrival of customers, which can mean checkouts being empty one moment, and having to deal with a long line the next. So to keep queues short and customers happy, supermarket managers will have to put up with seeing staff doing nothing. Alternatively, managers can slash staffing levels – and have customers complaining about the length of queues. Alas, you can't have happy customers and constantly occupied staff.

One way to deal with this dilemma is to set up one big queue in front of all the tills, a strategy often seen at banks or at theme parks, for example. The effect on waiting times can be dramatic: if you replace five queues in front of five tills by one queue feeding them all, the average waiting time will drop by a factor of five.

So given all this clever theory, why is it that the queue next to you so often finishes first? As Robert Matthews points out, even if your queue is the same length as both your neighbouring queues, as all three are equally likely to suffer from random delays, the chances that yours will suffer the least is just one in three: two-thirds of the time, one or other of your neighbours will do better, and finish before you. Matthews is interested in queuing as one of the manifestations of Murphy's Law – 'Anything that can go wrong, will go wrong' – on which he has published research. He has coined a Murphy's Law of queues: 'If your queue can be beaten by a neighbouring one, it will be.'

It is worth taking a short diversion to study another issue that has fascinated Robert Matthews – the seasonal search for socks. An average family of two adults and two children has two pairs of red socks to hang out each year for nuts, sweets and other gifts from Santa. However, as Matthews points out, this ritual highlights one of the most irritating problems faced in modern life – the odd-sock epidemic. Aunts and grandmothers are famous for giving socks at Christmas, and Matthews's research can also provide useful advice on what sort they should buy.

Anyone with a diverse collection of socks will have been struck by the proliferation of odd socks in drawers and will have spent some time every week in the quest for a complete pair. Fortunately, Robert Matthews has published a solution to this age-old problem in the journal *Mathematics Today*. He resorted to combinatorics, a mathematical discipline used to analyse combinations, arrangements and patterns. Inspiration struck, he says, 'When I was cleaning out the bath one morning.' After a frenzy of calculation, he uncovered three Murphy's Laws of Odd Socks.

1 'If odd socks can be created, they will be': when two socks go missing at random, it's far more likely that they will leave

behind two odd socks than conveniently disappear as a complete pair.

2 Random loss of just half the socks typically cuts the number of complete pairs left by three-quarters – and those that do remain will be lost in a sea of odd socks.

3 Even if you clear out all your odd socks, the problem of finding matching pairs still remains formidable: you will have to rummage through around one-third of ten pairs of socks to have a reasonable chance of finding a single matching pair.

'At first glance it may seem ludicrous to say that if odd socks can be created, they will be,' Matthews says. 'However, a moment's reflection reveals its plausibility.' Imagine a drawer containing only complete pairs of socks of different designs. The only assumption we need make for the analysis to proceed is that the loss process is random, with every sock as likely to go missing as any other. If one sock goes missing, it creates an odd sock, the partner of the sock you've just lost. When the next sock goes missing, it could be either that one odd sock just created, or a sock from a still-complete pair. As the latter outnumber the former, it is clearly more likely that another complete pair will be broken up, leading to the creation of yet another odd sock 'We can thus see glimmerings of evidence that Murphy's Law really does affect sock drawers,' Robert Matthews says.

Attempts to beat Murphy's Law of Odd Socks usually take the form of practical measures for keeping pairs of socks together, such as putting them into pillowcases before they go into the washing machine. Ideally, of course, we would like to beat the Law without having to go to such pains. The simplest solutions are to replace all our distinct pairs of socks with identical ones; to go the way of the elderly Albert Einstein and eschew them; or to show stoic indifference to odd socks.

'Happily, however, combinatoric analysis shows these dreary solutions are unnecessarily draconian: we can allow ourselves a little variety,' says Matthews. He recommends that people (or aunts seeking socks as presents) buy two types of sock: the red 'Christmas' socks and one other colour. Losing half the socks at random typically cuts the numbers of both types by half, and thus the number of possible pairs by three-quarters, as before. However, these remaining socks are not lost among a myriad of odd socks, and in general equal numbers of both types of sock will go missing. 'And of course you can guarantee that after pulling out just three socks, you can get a matching pair if you are in a hurry in the morning,' Matthews says. He admits, however, that the use of combinatorics has not answered the greatest mystery of all. Where do those missing socks go?

seven

Snow

Have you marked but the fall o' the snow
Before the soil hath smutched it?
O so white! O so soft! O so sweet is she!
Ben Jonson, *Celebration of Charis, IV* ('Her Triumph')

No Christmas would be complete without it. One-quarter of the globe, or 48 million square miles, is snow covered at some time during the year. Snow, locked into glaciers, permanently covers one-tenth of the planet's land area. An eiderdown of fresh snow falls each year on nearly one in every four square miles of dry land.

Snow can also decorate our language. It is a powerful metaphor, a symbol of purity, isolation and transience. Think of Snow White, for example, those ranks of doomed snowmen, or the winter scenes on seasonal cards. Yet snow is more than an essential ingredient of the Christmas experience. Watching every delicate flake, understanding it, forecasting it – all this is a major undertaking on which our society depends. World-wide, at least a third of the water used for irrigation comes from snow. Snow is big business, whether you are skiing, trying to keep automobiles moving in a major city, or even gambling on it.

Scientists study every aspect of the cycle that gives us snow, from the seeding of a single crystal high in the sky to secrets buried deep in the resulting snowpack, the layers of snow forming a 'time machine' that can be used to study climate change over hundreds of thousands of Christmases.

The Meteorological Office in Bracknell, England, hunts for

snowflakes with an aircraft called *Snoopy*, an ancient Lockheed C130 Hercules with a long instrumented snout. In Japan, researchers create artificial snowstorms in the laboratory. In the United States, snow studies are used to help track pollution or forecast how much irrigation will be available to farmers. Meanwhile, Germany's Fraunhofer Institute for Chemical Technology in Pfinztal has developed environmentally friendly forms of artificial snow. Some scientists are even beginning to find ways of manipulating snowfall. Meteorologists impregnate clouds with silver iodide to induce snowstorms, while ski resorts use bacteria to make snow.

THE SEASONS

Natural snowfall, when 'Frau Holle shakes her featherbed', as some Germans put it, depends on a combination of the Earth's orbit around the Sun and a 23.5-degree tilt in its axis. Together, the orbit and tilt conspire to give us our seasons, controlling the direction of the Sun's rays relative to the ground and the number of hours of sunlight each day.

Astronomers talk about equinoxes, when the length of the day and night are equal, and solstices, when the Sun is at its highest and lowest in the sky at midday. These times have been used to mark out the seasons. Thus, in the northern hemisphere, spring is deemed to start at the vernal equinox (near 21 March), summer at the summer solstice (near 21 June), autumn at the autumn equinox (near 21 September) and winter at the winter solstice (near 21 December). As the Earth sweeps around its orbit, the northern and southern hemispheres trade places in being closest to the Sun's rays. In June, the Earth's axis is tilted such that the northern hemisphere 'leans' towards the Sun. The Sun's rays strike the northern hemisphere more directly and for longer periods of the day. This is summer in the northern hemisphere. In December, the axis is tilted such that the southern hemisphere 'leans' towards the Sun. This is summer in the southern hemisphere and winter in the northern hemisphere.

The differences in the energy received from the Sun at any place on Earth are due to the Sun's height in the sky at midday and the associated length of the hours of sunlight. This period is longest at the summer solstice and shortest at the winter solstice. These days, however, neither solstice coincides with the hottest or coldest weather. Inertia in the Earth's response to the Sun's rays delays the warming and cooling of the hemispheres by several weeks. The Earth has a slightly elliptical – oval-shaped – orbit, so it is closer to the Sun during the northern hemisphere's winter. However, the variation in distance is so small – from 147 million to 152 million km (91 to 94 million miles) – that this seasonal influence is overwhelmed by that of the tilt and other factors. For example, although the southern hemisphere's winter occurs when the Sun is most distant, the larger mass of water in this hemisphere tends to make winters warmer than at equivalent northern latitudes.

THE STUFF OF SNOW

Snow is made from one of the most bizarre, complex and misunderstood of all fluids – water. We all take it for granted, but for scientists, water is a constant source of surprise. It is only by understanding this ubiquitous liquid that we can understand the science of snow.

Water is liquid at room temperature, whereas substances made of similarly sized molecules to those that form water – such as ammonia, methane or hydrogen sulphide – are all gases. The boiling point, melting point and heat-conducting abilities of water are also peculiar, being far higher than those of any comparable substance. For example, far more energy is necessary to boil a pint of water than to boil an equivalent amount of other liquids. The differences do not end there. Most substances shrink when you cool them, while ice takes up more space than water.

For the passengers on the *Titanic*, the fact that ice floats on water was bad news. But when it comes to living things in general, it is very good news. If ice was denser than water, our lakes would freeze from the bottom up. Fortunately for things that swim, paddle and wallow, an insulating skin of ice forms on a lake to protect the waters below so that they remain liquid. Water also has dazzling properties as a solvent. As whisky drinkers know, it mixes readily with alcohol. Sugar, salt and other minerals also dissolve easily. Water is an ideal medium to transport nutrients into cells; this is why the search for alien life is closely allied with the search for alien water.

All of water's strange properties can be understood in terms of its component molecules and their 'social' inclinations. In 1784, the English natural philosopher and chemist Henry Cavendish described the chemical composition of water, a combination of hydrogen and oxygen, marking the first step towards understanding its molecular structure and thus its peculiar properties. Hydrogen and oxygen are elements, two of the hundred or so basic chemical building blocks of matter. The smallest portion of an element that can exist is an atom, which consists of a central positively charged nucleus, surrounded by one or more diffuse shells of negatively charged electrons that cancel its charge. More than a billion atoms would sit on the full stop at the end of this sentence.

Atoms do not stick together in a haphazard way. Those of one particular element will combine only in certain ways with those of another. At its simplest, a link, or bond, between atoms in a molecule is due to their sharing a pair of electrons, providing a kind of negatively charged 'glue' to hold the positively charged nuclei together.

The special properties of water are the result of the way negative electrical charge is smeared over the molecule. We now know that each water molecule consists of one oxygen atom and two relatively tiny hydrogen atoms, arranged like the letter

V. The positively charged nuclei of these three atoms are 'glued' together by negatively charged electrons. However, the oxygen is so greedy for the attention of electrons that the two hydrogens are denuded of negative charge, leaving their positively charged nuclei exposed. That exposed charge is strongly attracted to other electrons, notably those in the oxygen atom in an adjacent water molecule.

Two pairs of electrons stick out from each oxygen atom like rabbit ears, serving as sticky sites for electron-starved hydrogens on neighbouring water molecules. The resulting so-called hydrogen bonds can strap each water molecule to four others. Vast networks of bonds are forged in this way at room temperature. The molecules link up to make a liquid, rather than moving independently, as in a gas. The self-same hydrogen bonds also make water appear blue because they absorb a little red from sunlight. When water freezes, these hydrogen bonds hold each water molecule apart, at bond's length, so the structure of the solid is more open and less dense than that of the liquid. This is why we should blame hydrogen bonds for the loss of the *Titanic*.

Hydrogen bonding also plays a part in a puzzle that baffled scientists for years. Why is it that ice has a slippery surface that allows us to ski, skate and slide so easily? The answer came recently from experiments by a team led by Peter Toennies of the Max Planck Institute for Fluid Dynamics in Göttingen, Germany. The scientists peppered ice with inert, low-energy helium atoms, which bounce off a surface so readily that they are exquisitely sensitive to the movement and arrangement of its topmost layer of molecules. The way that the helium atoms ricocheted from a single ice crystal cooled down to −243°C (−405.4°F) revealed that the tiny amount of available heat energy was able to stretch and compress hydrogen bonds between the water molecules at the surface by about 10 per cent. One of the team, physicist Alexei Glebov, explains that these findings suggest that at higher temperatures, such as those

found in ice rinks, the topmost water molecules are bestowed with enough mobility to break and remake hydrogen
bonds so that they can diffuse across the surface like marbles
on a tray. That is why ice remains slippery, even far below
freezing.

Hydrogen bonds explain other mysteries about ice. Under
atmospheric pressure and at temperatures below freezing, the
water molecules link together in larger networks whose fundamental building blocks are 'puckered' six-membered rings of
hexagonal symmetry. This same symmetry is preserved in
snowflakes. There have even been efforts to stick together
water molecules, one by one, to study the structure of water
and ice crystals. David Clary, a theoretical chemist at University
College London, and Jon Gregory of Cambridge University
simulated the process using computer programs and 'Quantum
Monte Carlo' methods, a random way to test the shapes
adopted by clusters of water molecules. Complementing this
work, Richard Saykally of the University of California in
Berkeley has created clusters of two, three and more water
molecules in beams that are cooled to near the lowest possible
temperature, The results of these experiments support David
Clary's predictions. 'We have worked on clusters of two, three,
four and five water molecules, all of which join up to form flat
rings,' says Clary. 'But when you add just one more molecule,
then things change dramatically. With six water molecules you
form a three-dimensional cage.' A star is born: the smallest
man-made snowflake.

Efforts to stick eight water molecules together have been
made at Purdue University, Indiana, by chemists Timothy
Zwier, Caleb Arrington, Christopher Gruenloh and Joel
Carney. Their experiments produced tiny ice cubes, which
came in two forms with the same mass and structure, but with
different arrangements of the hydrogen bonds that bound them
together. Though the hydrogen bonds in the top layer of each
cube were oriented in the same manner, those in the bottom

layer were arranged in two different ways, with the bonds facing either in the same direction or in the opposite direction to those in the top layer of the cube. 'These findings verify what theorists have predicted for years,' Zwier says, 'namely, that the eight water molecules preferentially form a cubic structure. They also provide the first evidence that, even in very small water clusters, water has the capacity to arrange its hydrogen bonds in several distinct orientations.'

You may be wondering how, in the name of Santa, Zwier and his co-workers could study these tiny cubes. They applied an infrared laser to excite the clusters, causing the hydrogen bonds in the tiny cubes to stretch and contract. By analysing the wavelengths of the resulting spectrum and comparing their results with computer calculations from another study carried out at the University of Pittsburgh, they were able to determine the molecular arrangement of the hydrogen bonds.

Since Timothy Zwier's two ice-cube structures are virtually identical in terms of energy, it would appear that the configuration that a particular cluster takes depends entirely on the collisions it undergoes while it is being made. This has profound consequences when it comes to the number of forms, or phases, of ice. 'Water has more solid phases than any other known pure substance because it can form phases that differ only in the orientations of the hydrogen bonds,' explains Zwier. No wonder every crystal is different! To date, around a dozen solid phases of water have been found, with a few more awaiting confirmation. For example, one was created using a special anvil, tipped with diamonds, to compress water to 6,000 times atmospheric pressure at temperatures between −10°C (14°F) and −50°C (−58°F) to create a form of ice that may exist on other planets in the solar system. In the past few months a new type of ice, dubbed 'Nebraska ice', has been created by squashing water flat. Xiao Zeng of the University of Nebraska-Lincoln and colleagues in Japan forced water to form a two-dimensional glass instead of three-dimensional ice crystals.

The achievement grew out of computer simulations that came up with the startling suggestion that water contracts rather than expands when it is frozen under extremes of pressure, temperature and confinement: 493 atmospheres at −40°C between two water-repellent plates spaced one billionth of a metre apart. In the computer model, the water froze into ice crystals with the hexagonal structure of ice in which each water molecule has a hydrogen bond with its four nearest neighbours – but differed from normal ice in that its crystals were flat and just two molecules thick.

SNOW BIRTH

For snow to fall outside the laboratory you first need clouds. These form when warm, moist air is cooled, either by being pushed up as it flows over a mountain or when a wedge of cold air noses underneath it. If the air is very clean it can reach high levels of saturation with moisture, indeed become 'supersaturated', before droplets of water appear. For droplets to form at lower levels, they require a nucleus to form on. These are called cloud condensation nuclei. As the air temperature declines, tiny drops (with a diameter of about $\frac{1}{100}$ of a millimetre) condense out of the water vapour. Viewed from the perspective of any humans walking below, these vast congregations of tiny droplets form the pillows, tails and streaks we call clouds.

Although the temperatures within a cloud are often well below freezing, the suspended water droplets will usually stay liquid. These droplets are 'supercooled' and will remain in the air as a cloud unless they cool to extreme temperatures of −40°C (−40°F) and below, when they freeze into tiny crystals called 'diamond dust'. One manifestation of ice crystals high in the sky are wispy filaments called cirrus clouds. Another are optical effects that result from the crystals behaving like prisms, and take the form of strange halos seen around the sun, or bright spots within this, called sun-dogs or parhelia.

For snow to form at temperatures higher than −40°C (−40°F), a special particle called an ice nucleus is required. This acts as a base for the crystal to form on. These 'snow seeds' are many and various. However, because there are so few ice-forming nuclei compared to the number of dust particles in the atmosphere, they are difficult to identify. Snow seeds include soil, dust, bits of volcanic ash, or even material of extraterrestrial origin. Each year some 40,000 tons of fine particles descend from the heavens. They are called Brownlee particles after Donald Brownlee, the scientist who first collected them in a U2 spy plane.

Compared to the number of droplets wafting around in a cloud, there are very few natural ice nuclei. The actual number depends on temperature. For example, at −20°C (−4°F), there is only around one nucleus per litre. You might think that, from this figure, we would know about the propensity of any cloud to form snow from the temperature alone. However, nature is never quite this straightforward, as Tom Lachlan-Cope, the British Antarctic Survey's 'snowman' explains. Many clouds at around −10 to −20°C (14 to −4°F) contain many more ice crystals than we would expect. These complex 'mixed-phase clouds' show that weather scientists still have some way to go to understand the physics of a seasonal sprinkling.

SNOW DIVERSITY

The snowflakes that float down from the sky differ from one another because of the way they attract new water molecules to their 'corners'. As the crystals fall through air of different temperatures and humidities, and are buffeted by winds, each grows in its unique way. No two snowflakes are alike because their shape arises from the interplay between the random arrival of water molecules and their preference for assembling in a hexagonal fashion. The crystals devour water vapour in the cloud, growing rapidly until they are large enough to fall as snowflakes, the so-called Bergeron Findeson process.

Large-scale flake growth was modelled using a computer in 1987 by Johann Nittmann of Dowell Schlumberger SA in France and Eugene Stanley of Boston University. They were building on work that dates back to at least 1949, when the Russian physicist G.P. Ivantsov devised an equation to predict the shape of a flake's tips. 'Each move to assemble a water molecule can be thought of as the spin of the wheel in Monte Carlo,' says Stanley. 'The sequence of, say, the trillion molecules needed to make a snowflake corresponds to a trillion spins of the wheel.' A snowflake tends to take on more molecules at its tips, as its growth screens off less accessible surfaces towards the middle of the flake. 'It is very difficult for a random walker [i.e., a randomly moving water molecule] to walk along a narrow fjord in a snowflake without sticking on the side wall,' says Nittmann.

Using computer simulation, the team re-created the random motions of up to 20,000 molecules. The results surprised the researchers. 'When we put the model on our computer we got a picture that so strikingly resembled a real snowflake we were flabbergasted. No one had ever used a computer to generate a picture that really looked like a snowflake,' says Stanley.

Atmospheric snow crystals come in infinite combinations of plates, stellars, columns, needles, spatial dendrites (in other words, branches), capped columns and irregular crystals. At around $-1°C$ (30.2°F), the ice crystals grow into thin plates. As the temperature drops to $-11°C$ (12.2°F), they grow into hollow columns. At $-15°C$ (5°F), everyone's idea of a conventional snowflake emerges, with star patterns starting to grow.

Stanley and Nittmann found that by tinkering with their computer model, they could make the two basic forms of snowflake: the flat hexagonal plates and the Christmassy variety with six feathery arms. The end result seemed to depend on the winds. With a great many random gusts, it becomes possible to 'fill in' the crystal so that hexagonal plates result. As ice crystals fall through the cloud, they clump together by a process called aggregation, which involves the slight thawing of crystals. These wet crystals then collide and freeze again, forming snowflakes. Aggregation works best at temperatures around freezing – too cold, and it does not occur because the crystals are too dry. The biggest snowflakes form when the temperature is between zero and 2°C (32 and 35.6°F). If the temperature is higher, the snow melts to form rain or sleet. This can make forecasting the weather tricky, since at high altitudes a melting snowflake can appear to be a fat water droplet to a ground-radar network. However, *Snoopy*, the geriatric research aircraft belonging to the British Meteorological Office, has studied ways of distinguishing the two.

The ice crystals swirling through clouds can have other effects. Their collisions, disintegration and interaction with water droplets can generate electrical charges that result in radio static and even in lightning discharges, known as St Elmo's fire. This is a silent, brushlike corona of atmospheric electricity that appears around tall objects and is often connected with the action of ice crystals, snowflakes and graupel, or soft hail.

SNOW OBSESSIVES

The snowflakes that adorn Christmas cards are idealized versions of the real thing. Most snowflakes are ugly sisters. Nonetheless, the underlying symmetry of snow has been apparent to our ancestors for millennia. The Chinese commented on it in 135 BC, and Europeans had noted it by the Middle Ages, when, in 1260, the Dominican philosopher, scientist and theologian Albertus Magnus wrote about snow crystals. At the beginning of the seventeenth century the same subject beguiled the German mathematician and astronomer Johann Kepler. 'There must be some definite cause,' he wrote in 1611, 'why, whenever snow begins to fall, its initial formation invariably displays the shape of a six-cornered starlet. For if it happens by chance, why do they not fall just as well with five corners or with seven?' In his pamphlet *On the Six-cornered Snowflake*, Kepler drew parallels with honeycombs and the pattern of seeds inside pomegranates, but he was unable to explain the flakes' hexagonal form. This was also studied by the English scientist and inventor Robert Hooke, who published drawings of snowflakes in his *Micrographia* (1665), an account of his microscopic investigations and one of the scientific masterpieces of the age.

The appearance of flakes was not widely appreciated, however, until the middle of the nineteenth century, when the book *Cloud Crystals*, with sketches by 'A Lady', was published in the United States. The lady in question had captured snowflakes on a black surface and then observed them with a magnifying glass. The end of that century saw perhaps the most famous snowflake obsessive of all. In 1880, at the age of fifteen, Wilson Alwyn ('Snowflake') Bentley of Jericho, Vermont, was presented with a microscope by his mother. By his seventeenth birthday, Bentley had accumulated some 300 drawings of snowflakes. His father encouraged his son's obsession by purchasing a camera, exactly what Bentley needed to preserve his fleeting snow crystals for ever. He made a successful photomicrograph of a snowflake on 15 January 1885. After sixteen years of collecting snowflakes at an average rate of seventy to seventy-five per storm, Bentley (who was a farmer by profession) had gained considerable knowledge and understanding of which storms, or parts of storms, and which temperatures yielded the best subjects. The year 1931 saw the culmination of his life-long passion when his book *Snow Crystals* was published. Around half of his 5,381 photographs were selected to illustrate it. On 23 December 1931 Bentley died of pneumonia. He had contracted the illness walking the six miles home in a snowstorm.

The legacy of all these photographs and drawings is the idea that no two snowflakes are identical. Yet a few years ago, Nancy Knight of the US National Center for Atmospheric

Research found twin snowflakes, matching hollow hexagonal prisms. Was the concept of each snowflake being unique a myth, after all? Although Knight's snowflakes looked identical, a precise match is so unlikely as to be impossible. Think of the numbers involved. Ice has a density of about 1 gram per cubic centimetre and from this, and the mass of a known number of water molecules, we can calculate that a 1cm cube of ice contains around 30,000,000,000,000,000,000,000 molecules, give or take a handful. A snowflake, then, might contain perhaps 100,000,000,000,000,000,000 water molecules. Given Tim Zwier's discovery that the eight molecules in his nanocubes can be arranged in two different ways, imagine the variety of ways that you could build a snowflake out of 100 million million million water molecules! The short answer is that you can't.

Aside from Nancy Knight's near-identical twins, there are many other exceptional flakes, notably in terms of their size. Most snowflakes are less than 1cm (½in) across. Given near-freezing temperatures, light winds and unstable, convective atmospheric conditions, much larger and irregular flakes can form. Then snowfalls deep and thick enough to impress King Wenceslas may occur. It is not known which snowflakes are the biggest ever to have fallen, but there are plenty of candidates. On 10 January 1915, huge snowflakes fell among the usual common-or-garden ones in Berlin. They measured up to 10cm (4in) across. 'Gigantic snowflakes!' the *Monthly Weather Review* of February 1915 trumpeted. 'They resembled a round or oval dish with its edges bent upward.' Monster flakes some 9cm (3½in) in diameter fell in Chepstow, England, in January 1887, according to a report in the journal *Nature* by a Mr E.J. Lowe. But the real mothers of all known flakes fell on Fort Keogh, Montana, in the same month as the Chepstow flakes. They landed on 28 January 1887 near a ranch belonging to Matt Coleman, and were described as 'larger than milk pans' in the *Monthly Weather Review*. They were said to measure 38cm (15in) wide by almost 20cm (8in) thick.

LYING SNOW

Even when snow has settled on the ground, there is no end to its dazzling diversity. 'Snow' is a common Indo-Germanic word (German *Schnee*, Latin *nix*, French *neige*) that is used for both falling and lying snow. Living in the frozen north, however, the Eskimos or Inuit are so familiar with snow that they have coined many different words to describe it. Or so the story goes. In fact, this has been called 'the Great Eskimo Vocabulary Hoax'. There are probably about a dozen words, similar to those you find in English (snow, sleet, slush, blizzard, avalanche, hail, hardpack, powder, flurry and dusting). The linguist Geoffrey Pullum speculates that the supposed number of Inuit words snowballed, so to speak, this century because it 'comports so well with the many other facets of their polysynthetic perversity: rubbing noses; lending their wives to strangers; eating raw seal blubber; throwing Grandma out to be eaten by polar bears.'

Whatever you want to call it, there is great variation in the water content of snow. Around 25cm (10in) of fresh snow can contain as little as 3mm (⅒in) of water or as much as 10cm (4in), depending on the snow's crystal structure, wind speed, temperature and other factors. We can also tell something about the nature of snow by the way it reflects sounds. A thick layer of fresh, fluffy snow readily absorbs sound waves. If the snow surface becomes smooth and hard as it ages, or has been polished by winds, it will help reflect the waves. Sounds may then seem clearer and travel farther.

The way snow looks is something of a puzzle. Snow is made up of ice yet there is one property that sets it apart from its component crystals – its colour. To explain why ice is blue and transparent, while snow is brilliant white, we need to understand how they behave in sunlight. Most natural materials absorb some light and this gives them their colour – blue in the case of water, for instance. Ice is also blue because it absorbs light in the red and yellow part of the sun's electromagnetic spectrum slightly more efficiently than in the blue part, so that more blue than red or yellow light passes through it. Because the variations in the absorption of the different colours of light are so small, it is only when ice is found in large quantities, such as in a glacier, that it appears blue. Glaciers are made even bluer by all the air bubbles, dust, dirt and small particles they contain. Light travelling through the glacier scatters and reflects off these imperfections, bouncing around inside the glacier. This gives the ice more time to absorb red/yellow light while deepening the blue.

In contrast, snow appears white. The complex structure of snow crystals results in countless tiny surfaces from which visible light is reflected. The light rays bounce from one crystal of ice to another, so that the light propagates randomly but efficiently through a snowflake before finding its way out again. No absorption takes place because the light waves travel only a short distance through the material due to the efficient scattering process.

The ability of snow to reflect light is also the reason why it is primarily melted by warm air rather than by sunshine. The same light-scattering process goes on in white paint, fog, clouds and the fluffy white beards of rubicund gentlemen riding on reindeer-drawn sleighs. When this scattering reaches a particular pitch of intensity, the light waves can interfere with each other in such a way that the light actually becomes trapped, a process called Anderson localization, which has been demonstrated in a powdered semiconductor by Diederik Wiersma of the European Laboratory for Non-linear Spectroscopy in Florence. 'You can think of it as "semi-frozen light" which keeps on running around in random loops,' he says.

COSMIC SNOWBALLS

The appearance of snow can be deceptive, however. One extraordinary example was recently snapped by a camera aboard NASA's polar satellite: a cosmic snowball that broke apart 15,000km (9,300 miles) above the Atlantic Ocean. Perhaps the biggest snowball ever seen, it was not made by enthusiastic schoolchildren but seemed to be an alien affair, one of the house-sized comets that pelt our planet every few seconds. There is, however, bitter scientific controversy over whether these giant snowballs really exist. The suggestion by Louis Frank of the University of Iowa that the Earth is the target for cosmic snowballs equivalent to about a million tons of water a day (thirty snowballs a minute, each weighing 20 tons) was – and is – regarded as preposterous by many scientists. At the time of writing, the great snowball debate still rages. Bashar Rizk of the Lunar and Planetary Laboratory in Tucson, Arizona, told a conference that 'Earth's sky would sparkle like a Christmas tree' if it was hit by a steady shower of such snowballs. His calculations suggest that, if it did occur, each resulting twinkle would be as bright as the moon and last for a minute.

There is another controversial theory that suggests that the Earth itself was once a giant snowball, cloaked in a white blanket measuring around a kilometre thick. The first hints of this worldwide freeze came in 1964, when Cambridge University geologist Brian Harland began puzzling over glacial debris close to the equator which suggested glaciers must have reached the tropics.

Around the same time, one of the first climate modellers, Mikhail Budyko of the Leningrad Geophysical Observatory, calculated that if the polar ice caps had spread past a crucial point, a runaway chill would have followed, eventually freezing over the whole of the planet.

The idea fascinated scientists, but the idea foundered because once the Earth was frozen there seemed to be no way out – the Earth would remain frozen. Then came a series of insights from Joe Kirschvink of Caltech, who came up with a solution – that volcanoes, protruding above the frozen landscape, would have carried on pumping out carbon dioxide, the greenhouse gas, even though the world had entered the deep freeze. On Snowball Earth there was no rain to wash this carbon dioxide out of the atmosphere. Instead it would have built up to higher and higher concentrations – until eventually it sparked off not just global warming but global meltdown. The Earth was roused from its cryogenic slumber.

A Harvard geologist called Paul Hoffman came up with further clues from carbon isotopes in rocks. Living things tend to boost levels of the carbon-13 isotope. At the time of Kirschvink's proposed snowball, Hoffman measured a gradual drop in carbon-13 relative to carbon-12 and reasoned that algae and photosynthesizing bacteria on the planet had struggled to cope with falling temperatures. A colleague, geochemist Dan Schrag, confirmed and extended his work.

Scientists are now starting to believe that this striking climate reversal happened as many as four times between 750 million and 580 million years ago. Each time, the Earth froze over completely for ten million years then warmed up rapidly when the ice melted because of the accumulated carbon dioxide. Life experienced a series of 'bottle-necks and flushes'. After the last of these climate shocks came the rise of the Ediacara – odd organisms that may well have been the first large multicelled animals. Hoffman and Schrag believe the snowballs were the key, creating conditions that would encourage new species to thrive, notably the first complex creatures on Earth. So we, Rudolph, Santa – and the rest of the planet's complex life forms – might owe our existence to the thaw of Snowball Earth.

SNOW HISTORY

Snowfall of a conventional kind can be put to intriguing uses. As snowflakes flutter through the atmosphere they scrub it of dirt. One study at the University of Wisconsin at Madison found that they are better than raindrops at gathering up pollutants. This might suggest that we should engineer snow-storms to scrub the skies of pollution. The scavenging proper-ties of snow are also handy for those interested in recording the chemistry of the atmosphere, providing clues to Christmas past and Christmas future. Fallen snow contains frozen relics of what was once in the sky. Slice through a layer and you see a series of snapshots of the atmospheric chemistry and pollutants, the particles of desert sand, volcanic ash, radioisotopes from nuclear tests and automobile pollution that the flakes have scoured from the skies. The same is true for ice sheets or glaci-ers, which are formed from the compression of snow over many years and contain the chemical fingerprints of gases, acids, pollen and dust. Evidence of ancient disasters is also trapped and frozen inside: traces of ammonium formate suggest forest fires; acid reveals volcanic eruptions; and radioactivity the fallout of nuclear tests and accidents.

One team of scientists from America and China has recon-structed a detailed climate record for the last 130,000 years from a 300m (1,000ft) core they drilled into the Guliya Ice Cap, a 200sq-km (77sq-mile) glacier sitting 6,700m (22,000ft) high in the mountains of western China. 'A record of this length from the sub-tropics is truly unprecedented,' says Ellen Mosley-Thompson, Professor of Geography at Ohio State University. The Ohio team cut their core sample into 34,800 pieces that were then tested for oxygen–isotope ratios, dust and pollen, and nitrate, chloride and sulphate ions. For years, researchers have assumed that the climate in the tropics and sub-tropics has been fairly stable, Mosley-Thompson says. But the new core from Guliya, along with other low–latitude ice-core records, suggests that these regions may have experienced considerable climate variability during the last 100,000 years.

In the remote interior of Greenland, another vast stretch of atmospheric history is trapped in the ice sheets. Between 1989 and 1996, the Greenland Ice Core Project, a thirty-strong team from eight European nations, drilled a hole through the icecap.

The highest point on the Greenland ice sheet was chosen as the site. This allowed the team to extract a core of ice that, if reassembled, would be close to 3.2km (2 miles) long, forming a record that stretches back more than 200,000 Christmases. The slow drilling process was described by team member John Moore: 'We drill about 2.5m [over 8ft] each time the drill goes down,' said the British glaciologist. 'It takes an hour to lower it down the hole, ten minutes to drill, and another fifty minutes to raise it.'

After initial analysis, the remainder of the 10cm- (4in-) diameter ice core was sent to Copenhagen for storage and more detailed analysis. The period covered by the compressed ice samples encompassed one ice age, an earlier warm period similar to the present, and part of the previous freeze. Physical and chemical analysis of about eighty dissolved constituents revealed the history of the climate. 'From the uppermost ice layers, we can see dust particles from the Chernobyl nuclear accident in 1986,' says Joergen Taageholt of the Danish Polar Centre. 'Much further down, we have found traces of acid rain in the ice caused by volcanic activity when Vesuvius erupted in AD 79.'

By crushing the ice, the breath of ancient atmospheres can be released. Trapped bubbles give evidence of concentrations of carbon dioxide and methane which may already have influenced patterns of climate change during the preceding millennia through the greenhouse effect. 'We are trying to look back in the past to understand how climate is controlled and to give us more confidence in predicting future climate,' says David Peel, a programme leader from the British Antarctic Survey.

Past temperatures can be deduced from the composition of oxygen isotopes. When ocean water evaporates in low and middle latitudes, the water molecules contain a given proportion of the two 'flavours' of oxygen, the heavy oxygen 18 and the light isotope, oxygen 16. But this proportion changes as clouds drift towards the poles. The heavier water molecules, which contain oxygen 18, condense faster than the light ones and precipitate sooner as rain or snow. By the time the remainder falls on Greenland, the levels of oxygen 18 are depleted. How much, of course, depends on the temperatures en route.

These ice records also contain hints that the climate may be more changeable than we think. This has profound implications for the long-term future of white Christmases. It seems logical to expect that the gradual build-up of atmospheric pollutants such as greenhouse gases will lead to an equally gradual change in climate patterns. But this comforting idea has been exploded by the climate records released from the ice, which revealed that during ice ages the global climate lurched from warm to cold and back again. Climatologists had reassured themselves that these 'climate flips' were linked to the ice ages and changes in ice sheets. In one projected scenario, armadas of icebergs were let loose and fresh water melted into the North Atlantic, disrupting the world's delicately balanced ocean-circulation system, which keeps Europe warmer than North America. Now, however, there is evidence that climate flips could occur in the absence of such ice sheets within a century or two – a mere eyeblink against the vast backdrop of geological time.

Evidence of climate flips locked up in the Greenland ice cores suggests that the Earth's climate cools significantly and abruptly in a naturally occurring 1,000- to 3,000-year cycle. Studies of isotopes of oxygen – an indirect measure of temperature – revealed these climate flickers during the final part of the last ice age, between 10,000 and 30,000 years ago. The temperature record shows fluctuations shaped like square waves, signifying a period of relative stability followed by a rapid rise or dip in temperature. However, the same record has been relatively flat over the past 10,000 years, suggesting that we can sleep soundly in the knowledge that climate flips occur only when the Earth is in the grip of ice. Or can we?

Evidence that these alarming flickers also took place long

after most of the ice sheets disappeared, and conditions closely resembled those of today, has come from work by Gerard Bond, a palaeoclimatologist at Columbia's Lamont-Doherty Earth Observatory in Palisades, New York State. Studying layers of rock fragments transported by glacial icebergs and sea ice to the North Atlantic, subsequently deposited on the sea floor then buried by sediments near Iceland and Greenland, he found that the abrupt coolings took place during the Holocene – the term geologists use to describe the past 10,500 years – after the last Ice Age ended and human civilization began to flourish.

The regularly spaced layers of debris showed that the amount of floating ice – and thus the global temperature – peaked every 1,000–3,000 years. Gerard Bond dated the peaks of this debris at 12,300, 10,800, 8,000, 5,700, 3,900, 2,750 and 800 years ago. 'The abrupt coolings in the Holocene era are not as great as those that occurred during the ice ages but still might be significant enough to cause severe winters, agricultural disruptions and other adverse impacts on people,' says Bond. The most recent cooling cycle might prove to be the Little Ice Age, which began around 1100 and peaked a few hundred years later. During this period, glaciers covered the Alps, Alaska, New Zealand and Sweden, snow blanketed Ethiopia's high mountains, global temperature was 1.1°C (2°F) lower than now, and Europe and North America suffered severe winters. 'If this is indeed a regular climate rhythm, it is still going on today,' says Gerard Bond. 'By understanding what causes these sudden climate changes when the Earth is relatively free of ice, we can anticipate the next event. The odds of a future climate jolt could be higher than we thought.'

But that leaves one huge puzzle: what causes these climate flips? One theory is that they are triggered by clock-like changes in the way the oceans circulate, the oceans being a crucial engine of climate, transporting vast amounts of heat energy around the globe. Another possibility is that the climate is responding to some external factor, perhaps the amount of radiation from the Sun beating down on the Earth. Recent findings have linked a 100,000-year climatic cycle to the influx of cosmic dust, the particles that help create snow in the first place.

THE CHRIST CHILD

Christmas itself is connected with one well-known climate cycle whose effects are felt every 3–7 years and seems to be getting more frequent as a result of global warming. For decades, scientists have known that, during this period, a huge region of warm surface water shifts towards the west coast of South America, affecting global rainfall and winds. The warming can be detected by a grid of buoys and sensors, called the Tropical Atmosphere Ocean Array, which measures atmospheric wind and ocean temperature. The popular name for this warming is El Niño – the Christ Child – as it usually coincides with Christmas. This benign name belies its impact, which is of biblical proportions. El Niño directly affects the climate of more than half the planet. Trade winds in the tropical Pacific Ocean slacken and reverse direction. A pool of warm surface waters shifts from one end of the equatorial Pacific to the other, accompanied by a great centre of tropical rainfall. On average, 10 per cent more rain falls than usual. In 1982, an intense El Niño was not predicted and was not even recognized during its early stages. Eventually, it caused thousands of deaths, and damage worldwide costing more than $13 billion.

The onset and intensity of Indian monsoons and African downpours are influenced by El Niño, as are the frequency, severity and paths of storms in the Pacific; the occurrence of regional droughts, forest fires, floods, mudslides and hurricanes; and the movement of shoals of anchovy, tuna, shrimp and other commercial fish. The latter is good news for some fishhermen, bad for others: a recent manifestation (1997–8) saw the loss of millions of adult sockeye salmon in the Bering Sea and large catches near San Francisco of the equatorial game fish mahi mahi.

El Niño also affects the jet stream across the United States, causing widespread flooding. Dramatic examples include the huge Mississippi River floods of 1993 and the destructive California rains in 1994–5. During the 1997–8 El Niño, the temperature rises in the Pacific Ocean off the Great Barrier Reef and Panama caused coral bleaching, when algae essential to the survival of the reefs were expelled. Nor is the Far East immune to the effects of El Niño, as underlined by the smogs that cloaked huge areas of South-east Asia when fires burnt out of control in late 1997/early 1998.

But if the Christ Child is second only to the seasons themselves in driving worldwide weather patterns, it also brings with it a gift: the prospect of long-term climate predictions. 'Our models cannot predict that it is going to rain the day after Christmas next year, but they can predict whether it will rain more than usual next December,' says Nicholas Graham, a climatologist at the Scripps Institution of Oceanography in La Jolla, California. 'We will not be able to say there will be a drought, but rather that there is a good probability of a drought occurring in a specific region of the world,' adds Mark Cane of Columbia University's Lamont-Doherty Earth Observatory. In 1986, Cane and colleague Stephen Zebiak put together the first computer model that successfully predicted El Niño. At about the same time, Graham and fellow Scripps scientist Tim Barnett developed a novel method that linked El Niño in the tropical Pacific to the global climate system. This advance allowed them to use Pacific sea-surface temperature information to predict rainfall, temperatures and other climate conditions elsewhere.

Almost a decade later, El Niño was used to predict corn harvests six months in advance and halfway around the globe. The link between El Niño and corn yield in Zimbabwe, where corn is the major food crop, was uncovered by Mark Cane and

Gidon Eshel of Lamont-Doherty and by Roger Buckland of the twelve-nation Southern African Development Community's Food Security Technical and Administrative Unit in Harare. The scientists showed that between 1973 and 1990, low rainfall and poor harvests fluctuated nearly in step with the periodic warming of eastern equatorial Pacific surface waters. Computer models that predict El Niños in August of one year can provide indications of probable crop yields the next April. This advance warning allows farmers to change strategies and gives policy makers time to conserve water, import food or take other precautions.

SNOW DREAMS

Given everything that scientists now know about the vagaries of the climate, the molecular structure of ice, the fractal structure of snow and the origins of flakes in clouds, can they engineer that white Christmas that we all dream of? Thanks to modern technology, it is now possible to summon snow without the help of the weather. In ski resorts during a mild winter, water is shot out through a cannon as a fine spray to produce snow. But this alone is not enough. As explained earlier, ice crystals need something to form onto – they don't just crystallize in thin air. Rather than relying on dust floating in the atmosphere, the snow–makers use harmless bacteria for the job. The resulting snow is not as good as the real thing, and tends to be lumpy.

The ski resort water-cannon method is not the only way of making artificial snow. Thanks to a facility built in Japan at the Suga Test Instruments Co. in downtown Shinjuku, snow can even fall on the hottest July day. Unlike the more primitive snow made for ski slopes, the $1.25-million computer-controlled Suga simulator facilitates the creation of natural snowflakes. The company specializes in the study of the weathering process, and sees artificial snow as a valuable supplement to the combinations of sunshine, wind pressure and rainfall used to put materials through their paces. Lab-made snow can also give insights into the atmospheric conditions that lead to the real thing.

Beyond this artificial snow there is, of course, the *really* artificial variety for those who want to create an off-season winter wonderland. The traditional plastic snow, of the kind made in Hollywood, consists of tiny flakes of white polyethylene that have to be collected and disposed of afterwards. Now there are alternatives. Sturm's Special, a manufacturer in Lake Geneva, Wisconsin, has developed biodegradable flakes of modified polylactic acid polymer, from corn or cheese by-products. The flakes are randomly sized, mostly irregular and plate-like.

Dietmar Voelkle and Frithjof Baumann of the Fraunhofer Institute for Chemical Technology in Pfinztal, Germany, have produced two new kinds of artificial snow. The first is biodegradable artificial snow, made of foamed potato and cornstarch. It is ideal for Christmas window displays, says its inventors, because it sticks to anything that is damp. 'One can even model icicles or build a snowman with the stuff,' Baumann enthuses. 'Add more water, and the flakes dissolve, so they can also be used outdoors without the need for sweeping up.' The second of their artificial snow creations is made of fire-resistant foamed polyethylene. Depending on how it is produced, it can be made to flutter and sparkle realistically for use in the theatre. This glitter snow has already made its stage debut in a performance of *Prince Igor* in the Badisches Staatstheater in Karlsruhe, saving the day when Hollywood suppliers were unable to deliver in time.

DREAMING OF AN ARTIFICIAL CHRISTMAS

There is as yet only one way to guarantee a genuine white Christmas: head for the snow line, the boundary at altitudes beyond which there is perpetual snow. In high polar latitudes, for example, the snow line is at sea level. In northern Scandinavia it is somewhere around 1,200m (4,000ft), in the Alps about 2,400m (8,000ft) and in the Himalayas about 4,500m (15,000ft). But given all that we know about snow, surely we no longer have to be slaves to geography and season? Surely it should even be possible to provoke the weather to provide that white Christmas on demand?

If we had the kind of resources that would make Croesus wince, meteorologists say that it should be possible to engineer the atmosphere to guarantee a good sprinkling. The trick is to reproduce the effect that triggers snow in the first place, when stationary cold air wedges under a moving mass of warm, moist air and lifts it so that snow condenses out. In Britain, for example, snow is produced in winter when an influx of warmer, moist air flows in from the west after a cold spell lasting for a few days.

Not entirely seriously, I asked the British Meteorological Office what I would have to do to make absolutely sure that it snowed on Christmas Day. In London, this is a rare event, occurring only once every twelve years or so. 'You would have to make Ireland much more mountainous – about the size of the Rockies – and extend it southwards to form a block in the Atlantic,' said a somewhat surprised spokesman. The resulting breeze from Europe would produce sufficient cooling over a week or so. 'Then you have got to make your block in the west fall over to let something back in, allowing weather systems from the Atlantic to ride over the cold air,' he continued. Somewhat unnecessarily, he pointed out that to erect and then dismantle a mountain range in Ireland made the scheme somewhat impractical.

Similarly hypothetical plans are under way in the United States to ensure that it snows from Georgia to New Jersey on Christmas Day. This jolly exercise is the work of Allen Riordan, co-ordinator of the Southeast Consortium for Severe Thunderstorms and Tornadoes, located at North Carolina State

University, who is an expert at forecasting severe weather. Riordan's first requirement was some serious manpower. The Army Corps of Engineers and the Canadian government would have to work together to steer the Santa Claus flow – a Siberian cold-air mass – over New England just in time to meet a warm and moist low-pressure system coming off the Gulf Stream.

Getting the cold-air mass to descend over New England instead of the Midwest, where it normally lingers, would be the most difficult part of the 'project'. The snow-makers would have to use a two-pronged approach. First, they would have to work around the clock to enhance the height of the Canadian Rockies, which should ensure a successful re-routing of the Santa Claus flow. 'It's going to take at least a couple of dump trucks,' Riordan says. To encourage the cold air farther east, the Canadian government would have to freeze Hudson Bay about a week before Christmas. The ice would maintain the cold air by keeping a high-pressure system on the surface of the land. Then, as the cold air headed southwards to the east of the Appalachian Mountains and over Virginia, it would meet warm, moist air coming over the offshore waters of the Gulf Stream.

To ensure that the warm and cold air masses collide, it is necessary to spark the formation of a low-pressure centre along their common boundary. To accomplish this, the US Navy would need to set off a series of massive explosions to create fireballs large enough to heat the air column and get the low-pressure system started. 'Then it will go by itself,' predicts Riordan. The warm air will head northwards over the Carolinas; cold air will push south under it, and snow should be underfoot on Christmas morning. Without such intervention, the real chances of snow in central North Carolina are 'very, very slim', according to the National Weather Service in Raleigh. By Allen Riordan's calculations, the best place for the chain of huge explosions is the Florida coast, near Jacksonville. A small price to pay for a white Christmas?

eight
Festive Fare

Be nice to yu turkeys dis christmas
Cos turkeys jus wanna hav fun
Turkeys are cool, turkeys are wicked
An every turkey has a Mum.
Benjamin Zephaniah, 'Talking Turkeys'

The first-known Christmas card was designed by the artist John Horsley in what the Victorians called the 'narrative style'. The side panels of the card depict acts of charity towards the less fortunate, a popular theme of the day. However, the subject of the centrepiece, surrounded by leafy trelliswork, is not a Dickensian righting of social wrongs but a scene with a great deal of contemporary resonance: a large family enjoying Christmas dinner.

To the scientist, seasonal cooking, whether roast turkey, potato latkes or stollen, is nothing more than a branch of applied chemistry The colour of turkey meat provides a glimpse of the bird's lifestyle, while scanners traditionally used to investigate the human body can be turned on food to reveal pockets of fat or a sixpence hidden in a plum pudding. A knowledge of thermodynamics can help us cook that same pudding to perfection. Indeed scientists have demonstrated that even the sauce you smother it with owes its consistency to a unique arrangement of molecules, in structures measuring up to 50 millionths of a metre across.

If you find that these materialist insights jar with the romance of Christmas tradition, don't forget that this tradition is the product of diverse cultural influences: when we settle down for our traditional Christmas dinner, we eat an Aztec bird by a German tree, followed by a pudding spiced with sub-tropical preserves.

Nature probably hones our appetite so that we build up reserves in preparation for winter. John de Castro of Georgia State University found that we tend to eat 10 per cent more calories between October and December and that when we eat in larger groups, as happens at Christmas, we tend to eat more and for longer. The food, naturally enough, varies from country to country, district to district, year to year and culture to culture. Traditional winter feasting has evolved into the cornucopia of seasonal fare on offer around the world today. In Spain, a soup of chicken stock, vegetables and pasta is eaten on Christmas Day. In France, a late supper, *le reveillon*, is held after Midnight Mass on Christmas Eve, varying according to regional culinary tradition: in Alsace, goose is the main course; in Burgundy it is turkey with chestnuts; while Parisians often feast upon oysters and pâté de foie gras. In Roman Catholic countries, fish is more usual Christmas fare than meat, notably on the holy fasting day of Christmas Eve. On Christmas Day, stuffed baked carp in a rich sauce is often eaten in Czechoslovakia and other places where this fish is caught. Pickled fish is consumed in Poland; eels and squid appear in some Italian towns; and the Swedes prepare a remarkable dish called 'lutefish' or *lutfisk*, fish that has been soaked for several days in lye. The alkaline water softens the tissue by dissolving its component proteins.

The diversity of Christmas cooking also reflects the availability of food. In Paris during the Prussian siege of 1870, the zoo and the sewers provided the Christmas dinner at Voisin's, a fashionable restaurant: consommé of elephant, braised kangaroo, antelope pâté, and a whole cat garnished with rats. Today, many Christmas revellers in the dry interior of southern Africa will tuck into the local harvest of mopane worms, the fat, spiny, mottled caterpillars of the emperor moth *Gonimbrasia belina*.

Recent research has found that in terms of protein, fat, vitamins and calories, the caterpillars compare favourably to meat and fish.

There are also, of course, many dishes associated with the other, non-Christian, celebrations that take place in December. For example, special varieties of fruit (*mazao*) and corn (*vibunzi*) are eaten by African-Americans in the United States during Kwanzaa, while Hanukkah fare includes latkes (potato pancakes) and *sufganiyot* (special doughnuts fried in oil).

The eating of ham at Christmas dates back to the sacrifice of a wild boar to the god Frey during a Scandinavian yuletide festival. Goose is another traditional Christmas meat, although today it has largely been superseded by turkey. Carp and salmon were popular in festive fish dishes, and sometimes both meat and fish were on offer. For example, in his *Diary of a Country Parson* (1773), James Woodforde describes a Christmas dinner that consisted of 'two fine codds boiled with fryed Souls round them and oyster sauce', followed by beef, pea soup, lamb, wild ducks, 'sallad' and mince pies. That evening, rabbit was eaten. This same spirit of excess is reflected in a rhyme by 'Whistlecraft' (the pseudonym of John Hookham Frere), dating from the early nineteenth century:

> They served up salmon, venison, and wild boars
> By hundreds, and by dozens, and by scores
> Hogsheads of honey, kilderkins of mustard,
> Muttons, and fatted beeves, and bacon swine
> Herons and bitterns, peacocks, swan and bustard
> Teal, mallard, pigeons, widgeons, and, in fine
> Plum puddings, pancakes, apple-pies, and custard …

A definitive catalogue of festive food can be found in *A Christmas Carol*, a book that conveys the sensuous appeal as well as the spiritual significance of Christmas: 'Heaped up on the floor, to form a kind of throne, were turkeys, geese, game, poultry, brawn, great joints of meat, sucking-pigs, long wreaths of sausages, mince-pies, plum-puddings, barrels of oysters, red-hot chestnuts, cherry-cheeked apples, juicy oranges, luscious pears, immense twelfth cakes [frosted cakes eaten on Twelfth Night], and seething bowls of punch.' After he read the book, the historian Thomas Carlyle, a man not given to impetuous gestures, dashed off an order for a large turkey. His wife remarked that he was seized 'with a perfect convulsion of hospitality, and has actually insisted on improvising two dinner parties with only a day between.'

THE TURKEYS INVADE

In Germany, Britain and elsewhere, the goose was for a long time the most popular Christmas dish. This was usurped by the turkey. This outlandish bird started its invasion of Europe around 1519, when ships first brought livestock back to Spain from Central America. The bird marched on to the Spanish Netherlands and then to England, where turkey farms were started in East Anglia. The eighteenth century saw great turkey drives from Norfolk to London. On Christmas Day 1815, the English essayist Charles Lamb wrote of the event as 'the savoury grand Norfolcian holocaust'.

In 1851, the turkey replaced the swan as the Christmas bird at Queen Victoria's dinner table. However, it was not until late-Victorian times that the turkey superseded the goose, or in the north of England roast beef, as the leading Christmas fare. (In the United States, being indigenous, wild turkey was already a general festive dish.) By 1900, *The Royal Magazine* estimated that in the London area alone 'The turkeys and geese cooked for Christmas would form an army, marching ten abreast, which would reach from London to Brighton', and the champagne used to wash them down would 'keep the Trafalgar Square fountains working incessantly for five days'. Once families had consumed what they could of the roast turkey, there followed cold turkey, and lots of it: minced and rissoled and devilled turkey for days and days to come.

The traditional turkey is soon likely to be altered by genetic engineering. Bernie Wentworth and his colleagues in the Department of Animal Science at the University of Wisconsin at Madison say that turkeys could lay around 20 per cent more eggs if they did not lapse into periods of broodiness. 'The aim of his research is to 'redesign' the birds so that they produce very little prolactin, the hormone that triggers broody behaviour. Wentworth has made an 'anti-sense' fragment of DNA that sticks to the gene responsible for prolactin, arresting its production. Biofundamentalists and others who oppose genetic engineering claim that such work will, in effect, reduce turkeys to machines that perform a market function. Whatever the ethics of designing creatures for human use, this is just one way in which DNA can be manipulated to influence the taste and texture of future Christmas meals.

THE TURKEY REVEALED

A chemist regards a turkey not as a tasty meal but as a combination of water, fat and protein, in proportions of around 60:20:20. Most of its meat is muscle tissue and consists largely of proteins. Two workhorse proteins – myosin and actin – make up the fibres that give the muscle its texture or grain. They lie in layers and slide past each other when the muscle is stimulated to contract. The colour of turkey meat holds a record of the bird's lifestyle. The difference between the white and dark meat is not down to its blood, or to the red oxygen-carrying haemoglobin, but to the closely related, oxygen-*storing* myoglobin, a molecule which, like haemoglobin, contains an iron atom at its heart.

There are two general classes of muscle fibre at work during exercise: fast and slow. Fast, or white, muscle fibres are used for rapid motion and rely on blood for their source of energy, while slow, 'red' muscle fibres are used for more sustained motion, fuelled by stores of fat. And that is where the red, oxygenated myoglobin comes in. Myoglobin retains the oxygen brought by blood until muscle cells need it. To some extent, the oxygen demand can be related to the body part's general level of activity: muscles that are exercised often and strenuously need more oxygen. The fast fibres are white because they do not need this red oxygen-storage molecule.

Turkeys do a great deal of standing around, but little if any flying, so their breast muscle is white and their legs are dark. The leg meat is also greasier than breast meat because fat is required to fuel their red muscle fibres. Game birds, on the other hand, spend more time on the wing and their breast meat may be as dark as their drumsticks, seasoned with myoglobin throughout.

Molecular makeup also affects the texture of meat. Turkey legs are tougher than the breasts because they are used more. As the bird grows and exercises, the muscles enlarge by increasing the number of actomyosin filaments within the muscle fibres: the more filaments there are to cut through, the tougher the meat. This is also why older birds make tougher eating. In general, however, well-exercised tough meat is tastier than tenderer, less-exercised meat. The tenderness and appearance of cooked meat also depend on the reaction of its component chemicals. At the molecular level, that means how the various components respond to the shake, rattle and roll of heat energy.

When myoglobin is heated during cooking it turns brown because the iron at its centre is oxidized, changing the way the molecule absorbs light. Proteins and other components break down into smaller fragments. Some are small and volatile, adding to our perception of the flavour. Others originate from a molecule called adenosine triphosphate (ATP), the chemical energy currency of cells. Of the vast cocktail of chemicals produced by cooking, the two most responsible for a meaty flavour are inosine monophosphate, which arises from the decomposition of ATP, and monosodium glutamate, the sodium salt of a naturally occurring amino acid.

When you roast a turkey, muscle fibres contract, until at temperatures above about 80°C (175°F) individual cells within the fibres begin to break up. At the same time, roasting snaps bonds within molecules – hydrogen bonds and so-called disulphide bridges – that maintain the shape of the proteins, which are coiled-up chains of amino acids. The proteins begin to unravel and adopt more open conformations so that the meat becomes tender. Continue roasting for too long, however, and the proteins build up a network of new chemical bonds. This crosslinking between proteins replaces the delicate association that previously existed, making the meat tough, a process called coagulation.

Another cause of toughness is the connective tissue that attaches the muscles to the bones. This contains three more proteins – collagen, reticulin and elastin. Neither reticulin nor elastin is much weakened by the heat of cooking, but the three strands that make up collagen unwind into separate ones, better known as gelatine, which is soft. The secret of an excellent roast, then, is to denature the collagenous tissues, while at the same time avoiding too much coagulation of the muscle proteins. Simple, really, Mrs Beeton.

Calculating the time required to cook a turkey demands skill, however. Denaturing and coagulation of proteins take place at different temperatures and rates, depending on their location in the turkey, whether in the legs, wings, breast or elsewhere. Fortunately, you can rely on a handful of general principles in order to cook the meat to perfection, according to Peter Barham, a physicist at the University of Bristol. The longer the bird remains at a high temperature, the more moisture it will lose and the greater chance there will be of muscle proteins coagulating. A minimum temperature of 70°C (160°F) will convert collagen to gelatin and ensure that the muscle fibres break down. Thus, at its simplest, the optimum cooking time can be defined as the minimum time required to heat the centre of the bird to 70°C (160°F).

TURKEY THERMODYNAMICS

The next problem – calculating how long the bird should stay in the oven – is one for thermodynamics. This well-established discipline took shape with the advent of steam power during the Industrial Revolution in Britain in the early nineteenth century. The name is formed from the Greek words meaning the 'movement of heat', and can be applied just as easily to turkeys as to steam engines.

In 1947, H.S. Carslaw and J.C. Jaeger contributed much to solving the seasonal problem of how long to roast a turkey in their reference work, regarded as the Bible of the field, *Conduction of Heat in Solids*. They came up with equations that govern heat transfer in a uniform sphere. Little did they realize that they had provided an excellent starting point for a theory to explain the correlation between the radius of the bird and the period it should spend in an oven. To ensure that the mathematics did not get out of hand, they made a few simplifying assumptions. In a Christmas context, these can be described as follows, says Peter Barham. First, the oven must be maintained at a constant temperature throughout. Second, the thermal diffusivity – a measure of how fast heat passes through the turkey – is independent of temperature and time. Third and most important of all, the turkey must be extremely plump, so much so as to be spherical.

The time taken by the heat to diffuse through the tissue so that the centre of this spherical turkey reaches a certain temperature is proportional to the square of the radius of the turkey. And by assuming that the spherical turkey has the same mass (M) as the real one, we can calculate its required radius. To do this we need to resort to a well-known formula which says that the mass of a sphere is proportional to the cube of its radius. By combining these two formulae, the radius disappears from the calculation and we are left with a cooking time (t) proportional to M to the power of ⅔. This would, of course, only hold if the turkey was a perfect sphere. Fortunately for the Christmas cook, Barham points out that the cookery books' simple algorithm of '20 minutes per pound plus 20 more minutes for a small turkey or 15 minutes per pound plus 15 minutes for a large bird' is, in fact, an equally good approximation for a real turkey. When it comes to the trimmings, you do not need higher mathematics to work out that, when you stuff the turkey, it is better to put sausage meat or whatever into the neck, where it will reach a higher temperature, than in the middle of the bird, where it might be undercooked.

The vegetables that accompany the turkey require us to think again about how molecular structure reacts to cooking. Potatoes contain granules of starch – a carbohydrate consisting of long chains of sugar molecules – which become soft, swollen and gel-like when cooked in water heated to between 58 and 66°C (136 and 150°F). At this point, the granules begin to soak up water so that they swell to many times their normal size. The perfect potato is full of such swollen, tender granules.

Hanukkah latkes are dollops of grated potato, mixed with eggs, matzo and salt, and fried in a well-oiled pan. As well as having symbolic significance (recalling the jar of oil that miraculously burnt for eight days and eight nights when Judas Maccabees rededicated a temple desecrated by the Syrian tyrant Antiochus), the oil has several important roles to play. It ensures that the mixture is in uniform contact with the heat source; it lubricates and prevents the latkes sticking to the pan; and it supplies some of the flavour by enabling browning reactions to take place. During these chemical reactions a carbohydrate unit from the potato starch reacts with the nitrogen-containing amine group on an amino acid, which may be free or part of a nearby protein molecule. An unstable intermediate structure is formed, which then undergoes further changes: this is called the Maillard reaction, and brown coloration and intense flavours are two of its many by-products. The trick is to brown the outside of the latke and ensure that the inside is done, since the moist interior remains largely water and never exceeds boiling point. Thus the potato has to be grated to ensure that there is no serious disparity between superficial and deep latke cooking time.

Roast potatoes also present challenges for the Christmas cook. When they emerge from the oven, they are often a disappointment, whether as a result of their anaemic appearance or the discovery that they look cooked on the outside but are hard inside. As with latkes, the problem is that we need to both cook and brown them, so that too high a frying temperature will brown the surface before the interior is cooked, and too low a temperature will cook the interior without any browning reactions to decorate and flavour the exterior. The roast-potato problem can be tackled by a two-stage cooking process. First, the potatoes should be boiled for a few minutes so that they acquire a gel-like surface layer. Then they should be basted with fat and put in the oven. During roasting, this surface layer prevents starch granules beneath from absorbing too much oil, while itself reaching about 160°C (320°F). The starch on the surface degrades and oxidizes to give the characteristic crispy brown coating.

Green vegetables require a quite different cooking philosophy from starchy ones such as potato and winter squashes. They need only the briefest weakening of their cell walls by heat and the extraction of a little water to make them tender. The bright-green colour that some vegetables develop a few seconds after being thrown into boiling water is a result of the escape of gases trapped in the spaces between cells. Their departure reveals the pigment (chlorophyll) that plants use to harvest sunlight. With a magnesium atom at its heart, the chlorophyll molecule absorbs both violet and red light so that reflected light appears green. Continue cooking the vegetables, however, and this magnesium atom is replaced by charged hydrogen atoms (protons), dulling the chlorophyll to the insipid green we all associate with overcooked Brussels sprouts.

HOW TO COOK THE TURKEY

Science can offer some interesting alternatives to the ovens, stoves and Agas that are usually deployed to cook the Christmas meal. To find out just how far it is possible to push back the frontiers of seasonal culinary art, I commissioned a study by Peter Barham and Len Fisher of Bristol University. Their brief: to devise an anything-but-traditional way to prepare the Christmas meal.

Anyone who has tried to cook the many components of the Christmas meal has probably discovered that the main problem to overcome is timing: different ingredients need cooking to different degrees, even different parts of the same food. When it comes to the turkey, for example, some cuts are more tender than others. While the wings and legs are loaded with connective tissue, the delicate breast meat contains much less and is inherently more tender and likely to become dry and tough.

What is needed is a method that will apply the right amount of heat to the right part of the turkey. The Bristol physicists decided that one way to fine-tune the cooking process would be to use lasers. The beams would be scanned over the surface of the bird so that the amount of heat applied at different places is constantly tuned to guarantee the very best possible results. And to do this, they proposed using a telescope. 'The equipment may not be available to everyone, but the result should be heavenly.'

This is not any old telescope but one that uses 'adaptive optics' that iron out the image imperfections caused by atmospheric fluctuations (the reason that stars twinkle). A computer-controlled system of sensors and precision pistons alters the position of each segment of a mirror – relative to its neighbours – to an accuracy of one thousandth of the thickness of a human hair – to compensate for atmospheric flickering.

The W.M. Keck Observatory of the University of Hawaii would be ideal for the job. The main reflector consists of 36 separate hexagonal segments each 1.8 metres across. By scanning laser beams across these mirrors and adjusting their focus and angle at the same time, it should be easy to steer the pinpoints of laser light to the required parts of the turkey mounted on a rotating spit.

The temperatures in different parts of the turkey would be monitored by inserting thermocouples (tiny temperature probes) to different depths. These measurements would be fed to a computer fitted with a neural network program, brain-like software. 'Such programs partly mimic the action of the human brain, in that (unlike some politicians) they learn from their mistakes and act to correct them', according to Barham and Fisher.

Using the thermocouple measurements, the neural net would fine-tune the position and power of the laser beams until the turkey is perfectly cooked. As a final touch, several of the beams would be boosted to full power and used to slice the turkey into pieces. 'We could even use the laser to weaken one side of the wishbone so that we would always win when it came to breaking the bone – then if only we could be sure our wishes would come true.'

Another set of lasers could be used to cook the vegetables. The Bristol team plumped for tuneable dye lasers, which allow the colour of the beam to be altered. Peas, for example, will absorb red light but not green light, so that a red laser beam will heat them up but a green laser beam will not. Using lasers we can apply just the right amount of heat to cook each vegetable to its own ideal temperature and at the same time avoid losing any water through breakage of cell walls, the cause of soggy vegetables. Each sprout, parsnip and carrot would of course be fitted with thermocouples, whose output would be passed to the neural network computer to optimize the laser wavelengths (colours) and hence the cooking rate.

After the first, and understandable, reaction of shock and horror, the astronomers who they approached agree that the proposal is novel and feasible – in principle, at least. 'Strangely enough, however, none have yet agreed to try it out. Perhaps the world isn't yet ready to bypass Dickens when it comes to the Christmas dinner. We are sure, though, that our time will come.'

THE DREADED BRUSSELS SPROUT

A member of the cabbage group, the sprout is thought to have been developed in northern Europe in about the fifth century, though the first clear record of it dates from as late as 1587. This vegetable is now part of the traditional Christmas meal, but nevertheless is greatly disliked, particularly by children. Science can offer a profound insight into the sprout paradox: the slightly bitter, sulphurous taste is the vegetable equivalent of a chemical weapon, and is meant to discourage would-be insect diners. Yet human consumers are not hurt in the crossfire (although there is some unpleasant fallout as stomach bacteria convert the sulphur to foul-smelling hydrogen sulphide). Indeed, for humans, sprouts and their close relatives are among the most nutritious of leafy vegetables, rich in minerals, fibre, protein, carotene and vitamin C. These plants, like many other fruits and vegetables, contain a range of 'non-nutrient' compounds that seem to protect against many types of cancer, including cancer of the breast, lung and colon, by reducing damage to DNA in cells.

Sprouts are also rich in one family of secondary compounds, called glucosinolates, notably sinigrin. Ian Johnson of the Institute of Food Research in Norwich, has discovered that sinigrin suppresses, at least under laboratory conditions, the development of damaged cells that may eventually develop into full-blown tumours. The protective chemical in sinigrin seems to be a volatile sulphur-containing breakdown product called allyl isothiocyanate, which is largely responsible for the smell and taste of sprouts.

Chemicals such as sinigrin are natural pesticides, meant to put off predators such as slugs, snails and rabbits. So how is it that we not only eat them with impunity, but also – hydrogen sulphide apart – benefit from the experience? The answer may lie in our evolutionary past. Thanks to our long history as omnivores, consuming a vast range of fruits and vegetables, we find ourselves today biologically adapted to an intake of plant toxins. Indeed, we seem to be so well adapted that our health suffers if we are deprived of a steady supply of the plants' chemical weapons. The aversion of children to such vegetables may be due to our Stone Age ancestry. Our bodies – and our appetites – evolved as a direct result of survival of the fittest, an endless arms race between our ancestors over millions of years as they fought off parasites, predators and disease. Physiologically, our natural feeding instincts still belong in the Stone Age, when the diet of hunter-gatherer man was very different from that of today, notably in terms of fat and salt content.

From the Darwinian perspective, our aversion to bitterness may have evolved to help us avoid poisonous plants, since many bitter natural products – for example, alkaloids such as strychnine – are poisonous, and the ability to detect them by taste was a necessary adaptation for survival. We can, of course, train ourselves to like bitter tastes, sometimes for good reason. The bitter principles contained in the quinine in a gin and tonic stimulate the production of saliva, making aperitifs such as this the ideal prelude to a meal.

In early childhood, then, we will all tend to dislike bitter tastes and thus vegetables such as brassicas. George Williams, Professor Emeritus of the State University of New York and

one of the pioneers of so-called Darwinian medicine, has suggested that children's feeding instincts are to look for the bland-
est foods and to avoid strong flavours because these are linked
to high levels of toxicity. At this critical stage in their development, children are better adapted than adults to avoid toxins.
Children also have a preference for sweetness, again reflecting
their innate anti-poison programming, but also ensuring that
they will adore breast milk, according to Dave Mela of the
Institute of Food Research in Reading, England. Moreover, a
taste for sweetness enables you to tell when fruit is ripe.

The nutritional value of vegetables is another reason why
children avoid them instinctively. Vegetables are not particularly energy-dense, explains Mela, and calories were a crucial con-
sideration to our ancestors thousands of years ago, given the
lack of four-wheeled transport and the constant need to hunt
food, fend off predators and so on. But the childhood aversion
to vegetables has serious consequences. One report from the
Cancer Research Campaign in Britain claimed that Christmas
dinner is now the only meal each year at which most British
children get an adequate intake of vegetables: 'Of the 300,000
people who get cancer each year in the UK, about a third of
cases are diet-related and potentially preventable. Getting children to eat vegetables is not a recent problem facing parents.
Repeated generations have tried to cope – often with limited
success. Much needs to be done to encourage children to eat
more vegetables.'

There is a further complication in this story. Some people
find the taste of sprouts, Campari or whatever, much more
bitter than others. This might not influence their food prefer-
ences when adult, since most of us learn to love bitter flavours
such as those found in coffee, tea and beer. As children,
however, these so-called 'supertasters' may have a particular
aversion to all bitter vegetables, not just sprouts. A pioneer of
supertaster studies, Linda Bartoshuk of Yale University School
of Medicine, believes that one-quarter of all Caucasians are

supertasters, the rest being 'nontasters', who make up 25 per cent, and 'mediumtasters'. Each group lives in a different 'taste world' so that sugary Christmas confections taste exceptionally sweet to supertasters, who also find the taste of caffeine and other bitters more marked. The burn of a chilli pepper may be twice as intense for supertasters as that experienced by non-tasters (who have little or no taste for certain substances).

Supertasters, mostly women, are usually distinguished because they find the chemical 6-n-propylthiouracil so bitter they cannot stand it. They can also be identified by examining the little projections on the tongue that contain taste buds. 'When someone sticks out their tongue, we can tell just by eyeballing it,' says Laurie Lucchina, one of the team at Yale. Supertasters can have up to thirty projections in a 6mm-(¼in-) diameter circle, while nontasters can have as few as ten projections in the same area, and the projections tend to be much larger.

The phenomenon is genetic. Last year, scientists found a family of around 80 'receptor' genes which code for proteins on the tongue that respond to bitter flavours. Because of the way the supertaster trait is inherited, if two mediumtasters marry, one-quarter of their children will be nontasters, one-quarter supertasters and half mediumtasters. Mediumtaster parents of a large family may not realize that this is why they may encounter particular aversion to sprouts from only some of their offspring. The same trait is also a cause for concern in the fight against cancer. As Adam Drewnowski of the University of Michigan explains: 'Supertasters may avoid cancer preventive compounds found in bitter and tart vegetables and fruit. Children who are supertasters will not like broccoli or Brussels sprouts, no matter what you do.'

This aversion was perhaps most famously expressed by George Bush Snr, who is probably a supertaster. While in the White House, he revealed: 'I do not like broccoli and I haven't liked it since I was a little kid and my mother made me eat it.

I'm the President of the United States and I'm not going to eat any more broccoli.'

'Many antioxidant flavonoids that are so important for cancer prevention are either bitter or occur in bitter-tasting vegetables and fruit,' continues Adam Drewnowski. For example, supertasters dislike naringin, an antioxidant that is the principal bitter ingredient in grapefruit juice and is being examined for its value in inhibiting cancer. This conflict between the health benefits of bitter chemicals and their off-putting taste creates a dilemma for producers and genetic engineers. Breed low levels of sinigrin into Brussels sprouts to produce the popular, milder-tasting variety and you risk reducing the health benefits, and perhaps also make the plants more vulnerable to certain pests.

GRAVY

Scientists are famous for asking profound questions, often questions that more humble folk could not even formulate. In the past year, such a question has been voiced by Len Fisher of Bristol University, and it is of deep interest to those who want to get the most out of Christmas Day: how much gravy should we pour on our Christmas lunch or dinner?

Dedicated to making science accessible, Fisher does this sort of project to show that science does not have to be high-tech to be fun. To investigate the gravy problem, he used a digital balance, sensitive to 0.01 grams, to measure the weight of Bisto gravy taken up by the traditional components of a roast. A series of experiments enabled him to establish a Gravy Uptake Rating scale, which can be described by a Gravy Equation, to solve this culinary mystery.

The equation shows gravy simply replaces the moisture lost during cooking. Roast potatoes, for example, lose around 30 per cent of their weight on cooking, and can take up a corresponding amount of gravy. Mashed potatoes only soak up a miserly five per cent. Fisher likens them to ground that has been saturated by heavy rain and can take no more moisture.

His studies reveal that potatoes are the main absorbers in the traditional Christmas meal, particularly when dunked in the gravy after being cut in half, while turkey meat does not absorb any at all. Carrots simply shrug off gravy, while peas take up an amazing 15 per cent of their own weight. However, there is a complication: some foods can take several minutes to soak up gravy. Potatoes can take up to 10 minutes. Parsnips take up their maximum amount within one minute, while peas, beans and brussels sprouts take up their maximum amount of gravy within 30 seconds. Fisher thus recommends that the turkey should first be eaten with parsnips and green vegetables, with roast potatoes last.

His research also revealed the surprising fact that it also matters how vegetables are prepared for best gravy absorption. Potatoes should be roasted (King Edwards are best) and parsnips should be cut lengthwise before roasting. The side of the vegetable which has been in contact with the roasting pan is the best when it comes to taking up gravy. And if you want to know just how much gravy to use, assuming you use a single slice of bread to mop up the last drop, Fisher recommends 60 grams (approximately 3 large tablespoons). This will cover the average dinner plate to a depth of 3mm. And the best breads to soak up any residual gravy? Fisher found the most porous of the commonly available breads is ciabatta bread (preferably slightly stale), which gave the most impressive results.

SCIENTISTS AND PUDDINGS

Plum pudding is a food fit for a monarch. King George I, sometimes called the 'Pudding King', tucked into one at six o'clock on 25 December 1714, his first Christmas in England. We still have the recipe: 2¼kg (5lb) finely shredded suet; ½kg (1lb) eggs; ½kg (1lb) dried plums, stoned and halved; mixed peel, cut in long strips; small raisins; sultanas; currants; sifted flour; sugar; brown breadcrumbs; 1 teaspoon mixed spice; half a grated nutmeg; 2 teaspoons salt; half a pint of 'new milk'; juice of half a lemon; 'a very large wineglassful brandy'.

This boiled confection of dried fruit, nuts, suet and so on, makes a striking appearance in *A Christmas Carol*: 'The pudding was out of the copper. A smell like a washing-day! That was the cloth. A smell like an eating-house and a pastrycook's next door to each other, with a laundress's next door to that! That was the pudding! In half a minute, Mrs Cratchit entered – flushed, but smiling proudly – with the pudding, like a speckled cannon-ball, so hard and firm, blazing in half of half-a-quartern of ignited brandy, and bedight with Christmas holly stuck into the top.'

Christmas pudding has its origins in frumenty, a type of porridge made from hulled wheat spiced and boiled in milk, which was sometimes used as a fasting dish on Christmas Eve, or as an accompaniment to meat. Over the years, eggs, mace, and dried prunes were added to it – indeed, meat was often thrown in. Some claim that the preparation of this meat dish merged with that of a great boiled sausage, called a hackin, an ancestor of the Scottish haggis. Somewhere down the years the hackin disappeared or amalgamated with the pudding, which became more solid and was boiled in a fine cloth like a sausage skin. The result was a pudding of the kind that Mrs Cratchit unveiled: a skinless relative of a haggis that is set alight, garnished with evergreen, and sometimes contains coins which – it is argued – are a lingering reminder of a Saturnalian tradition involving the drawing of lots. Today, scientists are perhaps the most determined consumers of plum puddings, which are sent by the British Antarctic Survey to its teams in Antarctica, whose members can sometimes find themselves up to 1,900km (1,200 miles) from the nearest base on Christmas Day.

In the past, the Antarctic scientists were given a 'Christmas box' containing probably the best-travelled Christmas puddings on the planet. Their journey started with a truck ride to the port of Grimsby, where they were loaded onto a research ship. Typically, it would sail at the end of October, via the Falkland Islands, to the Antarctic, arriving in early April. The boxes were then transferred by barge to the shore and carried by sledge to the Rothera base, a journey of up to 480km (300 miles), depending on pack-ice conditions (which affected where the ship moored). Finally, after more than eighteen months and 16,000km (10,000 miles), the puddings were consumed. The ritual had such significance that, in the sixties, a prominent mound in the Antarctic near the Survey's Halley base was christened Christmas Box Hill, after a group of scientists celebrated Christmas Day on it. Today Christmas boxes are no longer sent to Antarctica, and the puddings are part of the standard cargo of food dispatched from Grimsby in the autumn to be eaten at any time of year. However, they are still enjoyed on Christmas Day by the scientists 'holed up' in their tents while out on fieldwork, and by those in the British ice bases, Rothera, Halley, Bird Island and Signy.

The best puddings can take a long time to prepare. If they are left a year or two to mature, slow chemical processes take place, similar to those in a bottle of wine in a cellar. Then the pudding must be cooked. Like the roasting of the turkey, the thermodynamics of cooking Christmas pudding have been studied by physicists. Glenn Cox, a lecturer at Birmingham University, was surprised to find that cookbooks rely on rules of thumb rather than scientific principles to work out cooking times and temperatures. He therefore formulated an equation to describe the time required for the centre of a Christmas pudding to reach a given temperature. As was the case with the turkey, it was first necessary to make a number of simplifications so that the mathematical recipe was manageable: the pudding would be spherical and homogeneous (that is, currants, threepenny bits and so on were not taken into account). Cox also ignored the way heat-driven chemical reactions in the pudding affected the way it was cooked. Finally came his trick-iest task, finding the 'thermal diffusivity' of the pudding, the factor that determines the pudding's ability to store and transmit heat. Cox used a 'guesstimate', based on a comparison with substances such as salt, sand, beeswax, charcoal, water and alcohol.

To formulate his pudding equation, Glenn Cox resorted to complicated thermodynamic and mathematical equations, such as spherical Bessel functions and Fourier series. But what his equation does is simple: essentially it describes the temperature distribution in a sphere when its surface is held at a constant temperature. The equation, developed along similar lines to the turkey-cooking formula, revealed that the cooking time should vary in proportion to the square of the radius of the pudding, or to the volume to the power of two-thirds. Unfortunately, cookery books tend not to agree with these predictions. Cox found that many made no distinction between the cooking time of a 1-litre (1¾-pint) pudding and a that of a 2-litre (3½-pint) pudding, while his simple analysis of pudding thermo-dynamics showed that a 2-litre (3½-pint) pudding should be cooked for approximately 1.6 times longer than one that is half the volume. 'The proof of the pudding is in the heating,' he says. Of course, Cox is the first to admit that even a properly cooked pudding can taste awful.

CHRISTMAS SAUCES

One traditional way to cope with an unappealing pudding is to smother it with sauce – another rich seam of Christmas science. These liquids are of huge interest to surface scientists because of their complicated physical structures. Cream, brandy butter and brandy sauce owe their properties to a unique arrangement of molecules, in structures measuring up to 50 millionths of a metre. These toppings are all colloids, the name given to a sus-pension of another substance. They are important to the food industry because they affect the taste and texture of foods, and their shelf life. An emulsion is a colloid composed of two liquids, one (usually oil or fat) dispersed as droplets within another (usually water). Ice-cream, which is made up of solid fat particles, ice crystals and tiny air bubbles, is the ultimate colloid, because it contains solid, liquid and gas.

Cream consists of droplets of dairy fat, coated with a protein (casein) and dispersed in a watery solution of whey proteins. The concentration of the oil droplets dictates the creaminess, with double cream having a higher concentration than single. Brandy butter has a structure like that of cream that has been turned inside out, the butter being composed of droplets of water suspended in oil. Margaret Robins and Mary Parker at the Institute of Food Research in Norwich, England, have carried out an experiment that suggests that it is not possible to get the brandy into the suspended water droplets in butter without adding sugar. In a process that is not yet understood, the sugar helps the brandy pass through the surrounding fat layer.

Brandy sauce has a more complex structure. Like cream, it can be thought of as a suspension of materials in a watery solution. In this case, however, it owes its thickness to polymers, long chains of sugar molecules packed together in flour granules. The starch polymers come in two forms: one type, called amylose, consists of chains, while the other, called amylopectin, consists of branched molecules. Both types cling together within flour granules by means of hydrogen bonds. However, when the flour granules are mixed with hot water, the water molecules have enough energy to penetrate the granules so that their molecular structure begins to fall apart. As a result, these polymers are released into the watery solution of milk proteins, sugar, brandy and other components to make brandy sauce. The starch molecules cling to the water molecules and form an entangled mesh that captures the water-swollen granules. This dispersion of solid material (starch) in liquid (water) is known as a sol, and is what makes the brandy sauce thicken. Boiling, vigorous stirring or heating for a long time will thin this sauce, eventually causing the granules to disintegrate to leave a soup of finely dispersed starch fragments.

Tradition has it that chance takes over from the natural order of things at Christmastide, an idea that predates the festival of Saturnalia, when a priest, king or sacrifice was chosen by drawing lots. This notion has evolved into the custom of choosing a leader by hiding a small object in a cake or pudding. One example is the tradition of the Bean King, chosen on Twelfth Night in England, Holland, France and elsewhere by means of a bean hidden in the twelfth cake. In Denmark, a single almond was placed in the Christmas Eve rice dish. The practice of hiding a coin (formerly a sixpenny piece) in the Christmas pudding continues to this day.

This tradition gave me an excuse one Christmas to hunt the sixpence with the help of scientists at Cambridge University. Laurance Hall of the School of Clinical Medicine is the head of the team adapting magnetic resonance imaging (MRI), a non-surgical technique that provides images of the organs and biochemistry of living patients, for use on food. MRI scanners are invaluable tools that usually measure the distribution and behaviour in the body of sub-atomic particles called protons. The Cambridge scientists have used them to peer under orange peel, sausage skin and pie crust to measure the texture and composition of foods and estimate how they have aged, reacted to cooking and coped with different types of storage.

The key feature of magnetic resonance imaging is that it not only gives an image of a section through the food but also reveals the migration of fats and water, damage inflicted by freezing, and even the temperature distribution through a food as it is cooked in a microwave. The scanner can detect subtle changes, down to particles a tenth of a millimetre across, revealing whether a soggy pork pie results from excess water or fat in the crust. The movement of water and fat is highly significant for the texture of foods like chocolate biscuits, and in determining the reaction of breakfast cereal to milk. Magnetic resonance can even reveal bruising of fruits, such as apples, bananas and oranges, and help to develop transport methods that reduce damage. Would-be leaders will be pleased to learn that this technique can also be used to reveal a sixpence that has been slipped into a Christmas pudding.

CHOCOLATE

Christmas is also a time for eating chocolate – lots of it. But why do we think it is such a good idea to celebrate with fatty psychoactive food made from the seeds of the *Theobroma cacao* tree? Perhaps chocolate provokes 'psychopharmacological' reactions in the brain. Perhaps chocaholics 'self medicate' to overcome a dietary deficiency or lack of brain chemical. Perhaps there is a hormone link, since cravings often fluctuate around the menses. Perhaps a combination of these factors will ultimately explain why many drool at the mere thought of a truffle.

The ritual use of chocolate dates back long before Christmas to thousands of years ago, when it was first cultivated in the land between the Americas, including what today is Guatemala, Mexico, and Belize. Archaeologists tell us that chocolate was a key part of elite Mayan culture, notably in weddings. The Maya even had a special cacao god and other deities were often shown with piles of cacao pods and used the beans as money. But chocolate had a darker side. Cacao pods were likened to human hearts ripped out in sacrifice, and there was a strong symbolic link between the chocolate drink – often dyed red – and human blood.

The war-like Aztecs believed that drinking chocolate would bring wisdom, understanding and energy. Spanish conquistadores maintained that Emperor Montezuma consumed 50 cups a day because it was thought to raise more than his consciousness: its aphrodisiac properties would help him rise to the challenge of sex. No wonder chocolate has gone global in the 500 years since the Old World noted its use in the New.

Although chocolate quickly caught on as an elite treat in Europe, science first became aware of this food in the seventeenth century, according to Sandy Knapp of the Natural History Museum, London, the proud possessor of the first botanical specimen of a cacao plant to reach Britain.

The specimen was collected during a voyage to Jamaica by Sir Hans Sloane, a physician who collected enough objects in his lifetime to launch both the British and Natural History museums. Sloane's specimen was seen by the great Swedish naturalist, Linnaeus, who would go on to give it the Latin name *Theobroma cacao* – 'drink of the Gods.'

In Jamaica, Sir Hans noticed that the locals roasted these heavenly beans and turned them into a hellish brew, a dark, oily, bitter liquid (one theory has it that the name chocolate is derived from the Aztec or Nahuatl word xocalatl, meaning bitter water). He wanted to drink chocolate, believing it to have therapeutic properties, but found the flavour 'nauseous' (a taste of the original can be found today in Mexican mole sauce).

Sir Hans hit on the idea of boiling the crushed beans in milk in addition to using the vast quantities of sugar already used in chocolate preparation by European addicts. He sold his recipe to an apothecary in London, who marketed the product as 'Sir Hans Sloane's Milk chocolate'. Later the Cadbury family bought the recipe, and the rest is history.

Cocoa is a storehouse of natural minerals, containing more iron than almost any other vegetable and significant levels of copper, for example. Perhaps its rich magnesium content may explain the acute monthly cravings among pre-menstrual women, first noted in the 1950s: magnesium deficiency is known to exacerbate PMT. This is intriguing, though sceptics point out that cocoa powder capsules do not seem to relieve premenstrual craving as much as a bar of the dark stuff.

The stimulating effects of cocoa add to its appeal. Chocolate contains caffeine, but in modest quantities. There are also methylxanthine and theobromine, both caffeine-like substances. Although they do not act on the brain as strongly as caffeine, they do stimulate heart muscle, boosting the force of contraction and heart rate.

In the seventeenth century, a physician from Peru wrote: 'The cacao nut being made into confects, being eaten at night makes men to wake all night long, and is therefore good for soldiers who are on guard.' Indeed, some have suggested that it was Casanova's favourite bed-time drink to give him stallion-like stamina and keep his pecker up. Medical textbooks do note, however, that when taken in large quantities, these stimulants can induce nausea and vomiting. This can often be observed in children who have overindulged on Christmas Day.

More provocative is the discovery that chocolate contains a range of 'drugs' that could act on the brain, according to John Stein of Oxford University. 'No single one of them is present in particularly high concentrations in chocolate. But the combination may have an effect. It is subtle, which is why people are not addicted to chocolate.' One ingredient is phenylethyl-amine, a messenger chemical (neurotransmitter) naturally found in the brain that raises levels of another neurotransmitter, dopamine, in pleasure centres, boosting blood pressure and heart rate, while heightening sensation and blood glucose levels. This chemical relative of Ecstasy has been shown to produce a feeling of well-being and alertness, which may be why some people binge on the stuff after an upsetting experience, or perhaps to cope with the stress of Christmas shopping. However, much of the chemical is metabolized before it reaches the brain. Perhaps chocoholics are sensitive to small quantities and self-medicate because they have a faulty mechanism for controlling the body's natural levels.

More recently, it has been found that chocolate also contains substances that can act like cannabis on the brain, intensifying its other pleasurable effects. Three substances, from the N-acylethanolamine group of chemicals, can mimic the euphoric effects of cannabis, according to a study by Daniele Piomelli, Emmanuelle di Tomaso and Massimiliano Beltramo of the Neurosciences Institute, San Diego. Piomelli told me that he was first stimulated to look into the mood altering effects of chocolate when he became addicted to the stuff one grey winter in Paris. Now that he has moved to California, which is as sunny as his homeland of Italy, he is no longer a chocoholic. 'I limit myself to chocolate ice-cream.'

The work dates back to 1990, when scientists found a site in the brain which responds to cannabinoids, the class of compounds that includes the active ingredient of cannabis. Recently, they have discovered the substances in the brain that bind to this site. One is a fatty molecule dubbed anandamide, after the Sanskrit word for bliss. Piomelli investigated chocolate, which is rich in fat, because he suspected that it might also contain lipids related to anandamide. It turns out that it does. Scientists doubt that it is possible to consume enough chocolate for a 'high', even though it may be fun to try.

Studies on rats suggest that the wonderful taste of chocolate can boost the release of natural opiates – morphine-like chemicals – called endorphins, which are similar to those released by the brain at sexual climax and may contribute to the warm inner glow induced in susceptible chocoholics. This is backed by studies with the drug naloxone, which blocks the action of these opioids and reduces the intake of chocolate as a result.

Chocolate contains tryptophan too. An essential amino acid, its availability is the rate-limiting factor in the production of mood-modulating serotonin. Chocolate thus releases serotonin, another neurotransmitter with a calming effect and that plays a role in satiety. However, it is a puzzle why foods that are more effective at promoting tryptophan take up, such as rice and pasta, are less craved.

Even the smell of chocolate could be good for us. Pure cocoa has a complex spicy-milky aroma, not unlike a powdery nutmeg note, due in part to lactones (fragrant cyclic esters), thiazines (nutty smells) and pyrazines (roast). Research by Angela Clow of the University of Westminster suggests that, when compared with rotten meat, chocolate produces a larger amount of a powerful antibody called immunoglobulin A, which is secreted in the saliva and helps protect against colds. This is the first evidence that a smell can affect what is called the secretory immune system. No wonder the warm smell of chocolate is used in fragrances such as Angel by Thierry Mugler; and Rush by Gucci.

Another component of chocolate's appeal is the sensual way it melts, as chocolate body paint aficionados appreciate. Most fats and oils are complex mixtures of substances called triglycerides. But cocoa butter is uniform and so melts in a uniform way at around 34°C, just below body temperature. Chocolate draws heat from the tongue in the process of melting, creating a seductive sensation of coolness.

But while some claim chocolate's chemistry, precise melting point and neuropharmacology explain its attractions, others blame psychological factors. Nature has optimised us to crave sweet fatty food and chocolate represents the ideal combination of sweetness and fat. So much chocolate … so little time.

To investigate, experiments have been conducted at the University of Dundee by Marion Hetherington. 'Chocolate is the single most craved food' in studies of food cravings, she said. 'The best of foods, the worst of foods. Chocolate occupies a rather ambiguous position in the world of food, with both connotations of reward and high fat; unhealthy food; luxury and forbidden fruit; treat and temptation.' Two of her subjects banned chocolate from the house in a bid to curb their cravings but to no avail. One would cycle miles to a 24-hour-garage for a fix in the small hours, while the other would create her own substitute by mixing cocoa powder with sugar and butter.

Self-confessed chocaholics tend to be depressed, Hetherington discovered. A chocolate fix reduced cravings and hunger but had little effect on mood: they became slightly more contented but felt more guilty. In other words, although its chemical constituents may produce some effects our obsession with the brown stuff is more due to conditioning in early life.

We turn to it for comfort. Eating chocolate is pleasurable. It removes hunger and produces satiety, so it is associated with a feeling of wellbeing, said Hetherington. It also possesses powerful gustatory properties, such as the sensory gratification when chocolate melts and the fat is smeared around the mouth to air 120 different flavour volatiles, providing an instant hit.

We may crave chocolate because social norms say that we should eat it with restraint. 'We have a culture of using chocolate as a treat, as something special, as a mark of affection,' said Hetherington. It is an indulgence or 'forbidden fruit' rather than a staple part of the diet. On balance then, evidence suggests our seasonal craving for chocolate is more to do with psychological than neuropharmacological reasons. However, this is a subtle point: the fact that it is craved means it is affecting brain chemistry, albeit by indirect means, rather as tasting good food boosts endorphins.

Fighting the temptation to eat lots of chocolate at Christmas may be good for your soul but may cut your resolve. Indeed, it may even be hazardous to your mental health, according to a fascinating experiment by Ellen Bratslavsky and Roy Baumeister at Case Western Reserve University.

They tested students for their self-control, using one plate of chocolate biscuits and another of radishes. In the test, some students were told to eat radishes and not biscuits, while others were told to eat biscuits and not radishes before tackling an insoluble puzzle.

Students who resisted the temptation to eat the forbidden biscuits worked on the puzzle for an average of eight minutes before giving up. But those who had not had their self-control tested persisted for 20 minutes, said Bratslavsky. 'When people exert self-control – such as when under stress – self-control may fail in other spheres. Resisting temptation can be hazardous to your mental health.'

The good news is that chocolate is looking less hedonistic and more wholesome. Although feared because of its sugar and fat content, there is not much evidence that it causes obesity. It has been also blamed as a major cause of tooth decay. However, a study by Martin Curzon of the Leeds Dental Institute has concluded that there is little evidence for it being a 'cariogenic food.'

Decay is caused by acidity but chocolate does not reduce mouth pH (and thus raise acidity) as much as an equivalent sweet, or solution of sugar. 'Oral bacteria are less able to use the ingredients of chocolate to generate acid' he said. In milk chocolate, the dairy products speed the clearance of the confectionery from the mouth, compared with other sweets, giving bacteria less time to work in the mouth.

The chocolate–acne link is now thought to originate in folklore, having provided parents through the ages with an excuse to deprive their offspring of this indulgence at Christmas. Another fear is that chocolate would raise cholesterol levels. But although cocoa butter contains high levels of saturated fatty acids, studies show it does not boost cholesterol and thus the risk of heart disease. Chocolate could even help fight heart disease, according to Prof Catherine Rice-Evans of the Antioxidant Research Centre, King's College, London.

Recent studies have focused on flavonoids, chemical constituents of fruit and vegetables that mop up highly reactive chemical intermediates, called free radicals, that are implicated in heart disease, cognitive decline and ageing. There has been much excitement because one important family of flavonoids, the catechins, is present in greater amounts in dark chocolate than red wine, black tea, green tea, indeed anything else. However, there is a catch, since the chemical form of the catechins is different: unlike the other sources, chocolate's catechins are strung together to make polymers.

Chocolate has a high antioxidant activity, with one 50 gram bar of dark chocolate equivalent to two glasses of red wine. But little is yet known about the absorption, metabolism and uptake of these polymers in the body. We still don't know if the healthy flavonoids get in the right place, she said. More research is, as ever, needed. I, for one, would be happy to volunteer for these crucial experiments. What better excuse could you think of to indulge in this Christmas treat?

THE SMELL OF CHRISTMAS

There are many smells that we associate with Christmas and the effect of these aromas is more subtle than you may think. Among several food smells tested by Neil Martin, a psychologist at Middlesex University in London, chocolate was found to exert a significant calming effect on the brain. Martin asked forty volunteers to sit in a 'low-odour room', wearing goggles and headphones to block out other stimuli, while he wafted smells their way, using EEG (electroencephalography) to record their brain waves as they sniffed. Subjects were first exposed to the odours of real food: rotting pork, chocolate, coffee and baked beans. They then sniffed synthetic food aromas: almond, garlic and onion, strawberry, spearmint, vegetable, cumin – and chocolate. Only chocolate reduced brain activity, appearing to relax sniffers and make them inattentive. Chocolate connoisseurs might not be too surprised by this.

Smells can even help us carry out tasks, according to an experiment by Neil Martin's colleague Alison Gould. In one experiment, subjects completed a tedious 'visual vigilance task' in the presence of either no odour, an alerting smell (peppermint) or a relaxing one (bergamot). They did better with the peppermint. Another study showed that subjects tackled an unchallenging task more successfully when stimulated by an unpleasant smell – in this case, sour milk – while complex tasks were carried out more efficiently when subjects were soothed by a pleasant smell, such as that of an air freshener.

This kind of work is part of a growing body of evidence that suggests the rich smells of Christmas may affect us more than we realize. We become aware of different smells because inhaled air carries odour molecules, from the pyrazines in chocolate to cinnamon's cinnemaldehyde, to the roof of the nose. This is covered with a membrane containing sensitive, hair-like nerve fibres – the sensory receptors of the olfactory nerves that pass directly into the olfactory 'bulb' in the brain. As this area lies close to the brain centre responsible for emotions and cognitive behaviour, it is not surprising that smell is closely linked with emotions. A smell can transport people back in time, evoking a memory – for instance, of a childhood Christmas. In Marcel Proust's famous work *À la recherche du temps perdu*, the smell of lime tea and sponge cakes recalls childhood memories for the narrator. This kind of effect was observed in studies by Howard Ehrlichman and Jack Halpern of City University, New York, in which individuals were asked what memories were evoked by neutral words. Happier memories emerged when subjects were exposed to a pleasant odour (almond) than when they smelt an unpleasant one (pyridine).

This intimate link between the smells of Christmas – whether of roast Turkey, latke or a spiced cake from Nuremberg's famous winter fair – and our emotions was examined in experiments conducted by David Zald, a research fellow at the Veterans Affairs Medical Center in Minneapolis and at the University of Minnesota, working with José Pardo. Zald turned a brain scanner on a group of women as they sniffed a variety of smells. He chose to study women because they tend to experience smells more intensely than men. A dozen women were exposed to smells from plastic bags. Moderately bad smells includes garlic breath, natural gas and motor oil. The worst stench was that of a sulphur-bearing odour similar to rotting vegetables, bad eggs or a sewer. When the volunteers smelt this, a pair of almond-shaped complexes deep in their brains kicked into overdrive. Each half of the brain has one of these cell clumps, called an amygdala, and together they form a key part of the brain's machinery for creating emotional reactions.

Scientists have long known that there is an anatomical link between smells and emotions, a direct connection between the amygdala and the brain machinery that processes information from the nose. In the case of a bad smell, they can see this connection being used, as if the amygdala tells the rest of the brain, 'You really hate this stuff!' Pleasant smells, such as fruits, flowers, spices and presumably Christmassy aromas, evoke a weak response, and in the right-hand amygdala only. However, the context of the smell is very important, says José Pardo. For example, he explains, if the brain is supposed to evaluate the smell of burning wood, and the person is enjoying a cosy fire in a fireplace, 'probably the amygdala codes this as, "This is good, you can enjoy this", whereas if you're in the middle of a dark theater and you smell smoke, the message is "Fear, terrible, get out!"'

One of the dozen women studied by Zald and Pardo provided a good example of the influence of context. She found that the really bad odour wasn't too terrible, and brain scanning showed her amygdalae agreed. The reason: she had spent a summer holiday in Alaska near an oil refinery. 'It reminded her of that wonderful summer she had,' Pardo says.

The link between smell and emotion has been underlined by a study by Alan Hirsch, an odour expert at the Smell and Taste Research Foundation in Chicago. He found that male sexual arousal is strongly linked to food aromas. This discovery becomes somewhat perplexing when you consider that pizza and popcorn are more of a turn-on than the finest perfumes. 'The floral perfumes that we tested showed a median 3-percent increase in penile blood flow,' Hirsch reports. 'Cheese pizza caused a 5-per-cent increase, buttered popcorn caused a 9-per-cent increase, and a combination of lavender and pumpkin pie caused a 40-per-cent increase.' He even believes that smells can help you lose weight: 'The smell and taste senses work hand in glove. By saturating the sense of smell, you can fool the brain into thinking you've just eaten. You might think that giving them tempting smells would make them eat more. In fact, the reverse is true.'

This could come in handy after the Christmas bingeing, if one study is anything to go by. More than 3,000 overweight volunteers were given three vials to sniff, each containing a different food fragrance, and told to take three whiffs per nostril each time they felt hungry. They were advised to make no conscious changes to their diet or exercise. On average, the volunteers lost 2¼kg (5lb) a month for six months. The food odours tested for weight loss included banana, peppermint, green apple, cranberry, baked bread, barbecued meat, vanilla and cola.

AFTER THE FEAST

If you celebrate Christmas with a midday meal, you might well find yourself snoozing in the afternoon. In Britain, this is when the monarch's Christmas speech is broadcast, reinforcing the soporific effect of the meal. The drowsiness we experience after a heavy meal has been studied by many scientists, including Jim Horne, Director of the Sleep Research Laboratory at Loughborough University: 'We humans are designed to sleep twice a day, once at night and a short nap early afternoon, but in this part of the world we tend to repress that.' The nap is a remnant of the same primeval programming that makes all animals in the bush rest in the hot afternoon sun to avoid the heat. Hot environments increase the desire, and Horne notes that many cultures living near the Equator 'have conceded to the inevitable, where the afternoon siesta [a corrupted form of

the Latin *sexta hora*, meaning 'the sixth hour after wakefulness']
is the way of life.'

Jim Horne has also studied the contribution of alcohol to the
afternoon nap. 'The theory,' he says, 'is that if you are more
sleepy early afternoon, then it figures that alcohol will be more
potent. One would expect, then, that a pint of beer at
lunchtime has more effect than in the evening, when people
are more alert. Indeed, we find that it has about twice the
effect.' Alcohol interacts with the body clock to enhance afternoon sleepiness, so that one pint of beer at lunchtime is equivalent, in effect, to a quart in the evening. All of this carries a
serious implication. 'Drivers ought not to drink at all at
lunchtime and the legal blood-alcohol limit is no guide to
"safe" driving here,' says Horne. For those who wish to enjoy
the Queen's speech, or feel reinvigorated after a heavy lunch of
any kind, Horne recommends a cat nap. But this should be less
than fifteen minutes, 'Otherwise sleep really sets in and one can
wake up feeling very groggy and far sleepier than to begin
with.' Alternatively, mild exercise, a blast of cold air, a splash of
cold water on the face, or better still, a cup of strong coffee can
all help you to keep alert for the remaining Christmas
festivities.

nine

Christmas Spirit

Hot Tom and Jerry is an old-time drink that is once used by one and all in this country to celebrate Christmas with, and in fact it is once so popular that many people think Christmas is invented only to furnish an excuse for hot Tom and Jerry … Good Time Charley and I start making this Tom and Jerry early in the day, so as to be sure to have enough to last us over Christmas, and it is now along towards six o'clock, and our holiday spirit is practically one hundred percent.

Damon Runyon, *Furthermore*

The appearance of the very first Christmas card, designed by John Horsley, was marked by controversy. There was criticism that this seasonal message of goodwill actually encouraged degenerate behaviour. In one of its hand-coloured versions, it shows a girl in a green dress taking a gulp of red wine: a clear case of underage drinking. The protestations of the temperance movement had little effect, however, and by the time Christmas cards became popular, a couple of decades later, drunken celebrations were still regularly depicted. But nineteenth-century seasonal revellers did not need the encouragement of a festive greetings card to celebrate with alcohol. In 1894, the magazine *The Studio* sarcastically asked: 'If we investigated all the cases of drunkenness in all these years, could we find a single one remotely traceable to this design of Mr Horsley's?'

Even today, Christmas celebrations would not be the same without the attentions of the microbes that, through the process of fermentation, have provided us with alcohol for millennia. We all have a favourite microbial waste product. I like to sip a good claret, while some of my friends prefer to gulp down a cold beer in hot pursuit of the bottom of the glass. Whatever your approach to drinking alcohol, you are likely to consume more than usual at Christmas.

Scientists have now put together a detailed picture of the molecular events that take place within the body of the yuletide drinker, from the first sniff of a wine's delicate bouquet, to the absorption of alcohol into the bloodstream, to the resulting disruption of brain chemistry and the last groan of a hangover.

CHRISTMAS AND ALCOHOL

Whistlecraft's nineteenth-century rhyme describing the excesses of seasonal feasting, already quoted in Chapter 8, also provides an eye-watering, head-aching list of Christmas beverages: 'And therewithall they drank good Gascon wine,/With mead, and ale, and cider of our own,/For porter, punch and negus were not known.' Throughout the centuries, mead, ale and wine have been consumed in copious quantities during seasonal festivities. In Britain, for example, popular Christmas drinks included church ale, a strong brew, and lambswool, hot mulled beer with apples bobbing on its surface.' The latter was a shared wassail drink, of a kind that has forged bonds between revellers since Anglo-Saxon times.

The word 'wassail' comes from an Old English salutation ('Be well'); it denotes Christmas merry-making that involves carol singing and drinking the health of one's neighbours. The custom took various forms. In one New Year's Eve ritual, young women carried a wassail bowl of spiced ale about the local parish. They went from door to door, singing the following verse, presenting the drink to the inhabitants of any house they called on, expecting a small gratuity in return:

We have got a little purse
Of stretching leather skin
We want a little of your money
To line it well within.

In another variation, an empty wassail bowl was carried around by carol singers and a drink extorted at each door. The wassail bowl was also used in the home, sometimes with toast floating in it. This may be why we 'toast' someone, although other explanations of this term have been offered.

There are many such traditions. Ritual alcoholic abuse of the body has gone on for thousands of years, thanks to one of the most ancient processes of biotechnology – the fermentation of fruit and grain by the activity of fungi called yeasts. The result, both literally and figuratively, is a dizzying array of fluids, all of which contain alcohol (from the Arabic *al kohl*), also called ethyl alcohol or ethanol, the correct chemical name for this molecule.

The yeasts on grape skins, for example, give wine an alcoholic kick by fermenting the sugars in the juice. This biochemical process held attractions for the prehistoric people who discovered it: like cheese, wine is a partly spoiled food which resists further deterioration but which can be safely consumed. The alcohol produced by this spoiling has a preservative effect that can only be tolerated in small amounts by other organisms, including humans.

The action of yeast has fascinated scientists throughout the ages, spawning the study of microbiology and biochemistry. The first micro-organisms to be isolated in pure cultures were beer and wine yeasts. The Greek words for 'in yeast' gave us the word 'enzyme', meaning the proteins that cells use to form other molecules. This etymology seems very appropriate: by 700 BC, Homer's time, wine had become a staple beverage in Greece.

A BRIEF HISTORY OF BOOZE

Alcohol has been an integral part of seasonal celebrations – the planting of crops, harvests, the inundation of the Nile and so on – for thousands of years. If Christmas often seems to be marked by excessive drinking, remember that during the 'renewal festival' (*heb-sed*) for a pharaoh, feasting and drinking could go on for months. Nor were Neolithic people strangers to the misery of the morning after. Residues of liquid in an ancient pottery jar recently found in Hajji Firuz Tepe, a Neolithic village in the Zagros Mountains of Iran, show a wine similar to what we call retsina. The jar has been carbon dated to at least 5,400–5,000 BC, meaning that people have been drinking wine – and celebrating with it – for 2,000 years longer than previously thought. The yellowish residue at the bottom of the jar was analysed by a team led by Patrick McGovern of the University of Pennsylvania Museum, Philadelphia, using spectroscopic methods, which measured the light absorbed by the residue at various infrared and ultraviolet frequencies. It was found to contain the calcium salt of tartaric acid, which occurs naturally in large quantities only in grapes. The residue also contained terebinth tree resin, which was used in the ancient world as an additive to inhibit the growth of bacteria that convert wine to vinegar, and also as a medicinal agent.

Wine was used in early medical formulae, as well as recreationally, so the Hajji Firuz find provided insights into the pharmacopoeia available at that period in Egypt and Mesopotamia, and later in Greece and Rome. A second jar with a reddish deposit on its interior was also shown to contain a resinated wine. This may turn out to be the red to go with the white. Less clear is what the wine would have tasted like, though we can make an educated guess. The terebinth's lesser-known name is the turpentine tree. 'That's not exactly the kind of taste or smell that you'd necessarily want,' says McGovern. 'But I think it would dominate.'

Beer has also been available for seasonal celebrations for many thousands of years. Perhaps even before the grape was domesticated, ancient people had found that starchy grains – barley, wheat, millet and corn – could be treated so as to be fermentable. The trick is to allow the grain to sprout, when enzymes are released that chew up the starch molecules and break them down into their component chemical units, particularly glucose. In nature this action provides the sprouting seedling with an energy supply, but in brewing it supplies the glucose for yeasts to feed on.

Patrick McGovern's analysis of a fourth-millennium jar from Godin Tepe, another site in the Zagros Mountains of Iran, suggests that the brewing of beer may have been going on in the region at this time. We know for certain that barley and wheat beers were being produced in Egypt as long ago as the third millennium BC. Both bread and beer were consumed by all members of Egyptian society, supposedly even the gods, and at virtually every meal. It is a wonder the ancient Egyptians ever managed to build in straight lines. The conventional view of their brewing methods, based on records dating to around 1800 BC, suggested that malted grain was preserved by baking it into a flat bread, which was then soaked in water and fermented.

A few years ago, however, brewers from Edinburgh joined archaeologists in an attempt to uncover the beer-making secrets of the Egyptian pharaohs. The aim was to re-create beer as Tutankhamen might have drunk it, and the four-year project was undertaken by the Egypt Exploration Society and Scottish and Newcastle Breweries, led by Barry Kemp of Cambridge University. Expedition member Delwen Samuel set out to excavate Tutankhamen's brewery, and to find new evidence for ancient brewing using archaeological rather than traditional evidence such as hieroglyphics, tomb models, paintings and classical writings. The expedition's destination was Tel el Amarna, an ancient capital of Egypt halfway between Cairo and Luxor, and once home to Tutankhamen (who died at the age of eighteen in about 1340 BC), his father, King Akhenaton, and Queen Nefertiti.

Delwen Samuel believes that previous work has misinterpreted the ancient recipes suggested by paintings of the period. Instead, she examined actual samples of bread and beer found in and around the tombs, and especially in domestic areas where everyday beer was made. She was able to use the solid residues of beer left after evaporation, most of which were recovered from rubbish dumps where vessels had been discarded after breakage. Charred grains of emmer wheat, a very rare species still used in a few Mediterranean countries, and barley, were among the bread moulds and broken brewing jars found in the remains of the temple bakery-cum-brewery. Using microscopy, Samuel found that funerary bread was made mostly from emmer wheat, occasionally flavoured with figs, dates or coriander, while the grain used for beer was more varied – frequently barley, but also emmer and sometimes a mixture of the two.

Because of the remarkable preservation of some of the finds, Delwen Samuel was able to analyse the structure of starch granules in the remains of bread and beer, which provided important clues to baking and brewing methods. Her conclusion: the ancient Egyptians used a sophisticated two-step process, resembling brewing methods found today in Africa. The grain was divided into two batches. One was made into malt by the sprouting and gentle-drying method, then coarsely ground. The other batch may or may not have been malted, but was coarsely ground and heated in water. The two batches were then mixed together, allowing the active enzymes from the uncooked malt to break down the starch in the cooked grain, which would be much more susceptible to enzyme attack. This is an excellent system for producing large quantities of sugar from grain without the need for thermometers or other means to control the process. At this point, the chaff, which was still in the mixture, was removed by sieving, and the resulting sweet cloudy liquid was inoculated with yeast and fermented. 'The ancient Egyptians were much more sophisticated brewers than has been thought,' Samuel concludes.

The National Institute for Agricultural Botany in Cambridge grew emmer for the next step of the project: the re-creation of Egyptian baking and brewing recipes. The bread was, says Delwen Samuel, very palatable, with a 'distinctive, nut-like flavour'. The beer, made by Scottish and Newcastle Breweries using modern equipment, turned out to be much stronger and spicier than the modern equivalent. It was put on sale in London at Harrods department store for £50 a bottle under the name 'Tutankhamen's Ale'.

In societies where both wine and beer have been available, beer has always been the drink of the commoner and wine that of the elite. This may also explain why people are more inclined to celebrate a special occasion like Christmas with wine. The food-science writer Harold McGee points out that grain, the raw material for beer, is cheaper than grapes, can be stored for a long time in raw form before being brewed, and is easier and quicker to ferment so that 'To the Greeks and Romans, beer remained an imitation wine made by barbarians who did not cultivate the grape.'

FROM LIP TO BLOOD AND BRAIN

Exactly what happens when we drink wine depends on an individual's genes and body weight. However, the following is a reasonable guide to what goes on in the body of anyone who enters the seasonal fray with gusto. After a couple of glasses of wine, alcohol produces a general feeling of wellbeing. Another glass, and inhibitions may begin to crumble. Once a bottle has been consumed, you feel unsteady on your feet. Two bottles, and you are drowsy and confused. Three produce a drunken stupor and four can make an average man dead drunk, even dead if consumed quickly enough.

With the first sip, the bouquet, whether of King Tut's tipple or a fine Burgundy, is distinguished by the combination of taste with the more discriminating sense of smell that detects volatile vapours. These arise during fermentation, when not only alcohol but also several hundred other substances are produced. When a glass of wine or beer is raised to the lips, about 10 trillion odorous molecules get up your nose and waft over a yellow membrane that is plugged directly into the brain. For many years biologists puzzled over why we can smell anything at all: how could the water-repellent odour molecules pass through the moist cell walls in the membrane? Then a team at the Johns Hopkins Medical School in Baltimore, Maryland, found a family of proteins that transport smells into the cells. Once inside, each chemical component – in this case, those of the wine vapour – locks onto molecular docking sites, sending a signal to the brain that can, for instance, identify a wine as coming from the sunny side of a Burgundy valley.

Scientists used to think that the smell of these chemical components depended on their shape, which in turn affected how the molecules docked with the 'smell' sites located within the nose. But Luca Turin of University College London believes the *movement* of the molecules is more significant, reviving an idea that dates back to the work of the British chemist Sir Malcolm Dyson six decades ago. Molecules with very similar shapes can have different vibrations and thus different smells, Turin says. One molecule, shaped like a hamburger, smells quite different depending on whether the 'meat' is an iron or a nickel atom, for example. Similarly, boron and sulphur compounds of different shapes can smell the same because sulphur-hydrogen bonds and boron-hydrogen bonds vibrate at the same frequency.

Perhaps the strongest evidence that we smell vibrations has come from observing the effect in the molecule acetophenone of the exchange of hydrogen for its heavy brother, deuterium.

The molecule maintains the same shape, but the new vibrational frequency of the carbon–deuterium bond is lower than that of the bond with the lighter hydrogen atom. As a result, acetophenone and its deuterated version smell different.

When the hundreds of different vibrating molecules in the air interact with receptor cells in the nose, their signals travel first to the brain's 'olfactory bulb' for preliminary processing and then to the cerebral cortex, which interprets the pattern as a particular smell. How do we recognize more than 10,000 different odours? If each nerve carried a receptor tailored to a specific odorant (smell molecule), the brain would know what it was smelling simply by knowing which nerves were active. But to give the full gamut of aromas, from cinnamon to roast turkey, there would need to be as many types of receptor as there are smells. On the other hand, there could be just a few types of receptor, each reacting in a slightly different way to a given smell. The brain would have to compare lots of messages from this handful of receptor types to work out what it was smelling. Our eyes work in a similar way: they have only three types of colour receptors. By comparing the responses of red, green and blue receptors, the brain paints relatively few nerve signals into all the hues of the spectrum.

The first analysis of the human genetic sequence, published by rival teams in 2001, revealed our complement of odour receptors and how we have lost much of our sense of smell. This depends on a huge number of different receptor proteins, each tuned to a different sort of chemical odour molecule, which stick out from nerve cells in the neural cavity to dock with airborne molecules. The production of these proteins requires, in turn, a large number of genes. In the human genetic sequence there are more than 900 stretches of DNA that look, at first sight, like genes for these smell receptors. But the DNA shows our sense of smell, though important, is by no means as important as it was to our ancestors: scrutiny of those

900-odd 'genes' shows that 60% of them are now broken in ways that mean no protein can be copied from them.

Work by Stuart Firestein and colleagues at Columbia University has even started to link the receptors with particular smells. The first aroma to be matched was octanal, which smells like meat to some and has a light citrus odour to others. A good fit between octanal and smell receptor changes the shape of the receptor, triggering a cascade of processes that send information to the brain. That still leaves the problem of how these receptors' response to the complex mixture of chemicals in an odour is translated into a characteristic pattern of brain activity. Enter the bee. 'What we can smell, honeybees can smell. There is not a big difference,' says German scientist Randolf Menzel. Together with Jasdan Joerges and colleagues from the Free University of Berlin, Menzel conducted tests to find out what was going on in the insect's brain when it sniffed a carnation, whose smell has about thirty components. In this way, they cracked the code the brain uses to represent each smell.

The researchers studied the activity in the bee's olfactory bulb – the first stage in the smell pathway – which looks like a bunch of grapes, each 'grape' consisting of bundles of nerves called glomeruli. Their work showed that odours set up patterns of activity in the glomeruli: the more intense the smell, the greater the contrast in activity. The smell of a carnation was not simply a sum of the patterns made by each of its thirty component smells; the pattern is 'predominantly a simple addition but has subtle differences,' says Randolf Menzel. 'If you mix more than two components, say three or four, then a unique pattern develops.' Something similar is going on when we nose a wine. It should be no surprise, therefore, that different vintages of the same wine can create different patterns of brain activity.

Taste is crude by comparison with our sense of smell. Called chemoreception by scientists, it takes place in 9,000 taste buds on the tongue. The food or drink molecules that evoke taste are called sapid (from the Latin *sapere*, meaning 'to taste'), and in general they have to be soluble in water to penetrate the taste buds. Five types of taste bud respond to the saporous units in a drink: sweet, salty, sour, bitter and 'umami' (meaty/ monosodium glutamate).

Once swallowed, the first gulp of a drink passes down the gullet into the stomach, where it can be absorbed into the bloodstream through the stomach wall, a mucous membrane. The alcohol is taken up at different rates, depending on the drink; a warming toddy, such as *glogg* or *glühwein* (mulled wine), will intoxicate more rapidly than a chilled wine because the heat stimulates the blood supply to the mucous membrane, allowing more alcohol into the bloodstream. Sparkling alcoholic drinks, such as gin and tonic, are also more intoxicating. Carbon dioxide bubbles speed up the passage of the alcohol into the small intestine, where it is absorbed about three times faster than in the stomach; this is why champagne 'goes to one's head'.

The process within the brain that causes intoxication was discovered in 1996 by researchers at Washington University School of Medicine in St Louis, Missouri. A compound produced by alcohol-soaked brain cells – a fatty-acid ethyl ester – inhibits the release of messenger chemicals in the brain of a Christmas reveller, disrupting communication between nerve cells. This leads to slurred speech, clumsiness, slow reflexes, loss of inhibitions and short-term memory loss. 'I'm hopeful that we can now figure out a thousand-year mystery,' says Richard Gross, one of the Washington team, which also included Rose Gubitosi-Klug. 'Despite the structural simplicity of alcohol, nobody understands the biochemical mechanisms that are responsible for its neurological effects. It is a chemical in the process, not the ethanol itself, that is the active agent.'

While biochemists continue to investigate the effects of alcohol on individual cells, body scanners have revealed the havoc that an alcoholic drink can wreak across regions of the brain. One such study was carried out by Malcolm Cooper, John Metz and colleagues at the University of Chicago, using a scanning method called Positron Emission Tomography, in which a glucose molecule, labelled with radioactive fluorine, highlights the brain locations that have the greatest hunger for energy and thus the highest metabolic activity. The researchers found metabolic differences between the brains of those who enjoyed alcohol and those who did not; the former had more activity in the left side, and in particular the frontal and temporal lobes. The scanner also revealed a spectrum of effects that will, perhaps, not be surprising to anyone lurching out of a Christmas party: a surge of metabolic activity in the speech area, suggesting why drunks talk too much, and a surge in regions of the cerebellum that coordinate movement, which is why alcohol makes us stagger. Increased activity was also recorded in the limbic system, a region that controls primitive responses such as sexual arousal and violence. This might be related to 'brewer's droop' and boisterous behaviour.

The young are likely to be able to consume more alcohol than older people, and to become more rowdy when doing so, because alcohol is less likely to make them feel sleepy. That is, if young people behave anything like young rats. Junior rodents are more sensitive to alcohol-induced learning and memory deficits than adults. At the same time, alcohol does not make them feel as sleepy, which potentially allows them to drink more, according to studies by Scott Swartzwelder and colleagues at Duke University Medical Center and the Durham Veterans' Association Medical Center in North Carolina. 'The developing brain has exactly the wrong combination of sensitivities to alcohol,' Swartzwelder says. 'It's capable of staying awake through a prolonged bout of drinking, but at the same time, it is sustaining far more damage to memory and learning systems than an adult brain receiving an equivalent amount.'

Swartzwelder's studies suggest that as few as two drinks could inhibit learning and memory in a young person but would have little effect on an adult. At no other time can the brain absorb and retain so much information as during child-hood, and alcohol potently depresses that ability. A follow-up study – on 21–30 year olds for legal reasons – corroborates this theory. 'The memory performance while on alcohol was better among the older people in the sample than in the younger ones,' says Swartzwelder. Perhaps this is one reason why Christmas memories from childhood are more vivid than those from adolescence, when many begin to experiment with alcohol for the first time.

The passage of alcohol through the body is well understood. The body will treat it like a poison and attempt to break it down in the liver. This organ can deal with large amounts of alcohol, but it takes time – drink a litre of spirits in one go and you may die; pace yourself, and you can increase your tolerance to seasonal celebrations. A regular drinker will have more enzymes for breaking down alcohol than the occasional tippler.

Absorption of one half-pint of beer or one glass of wine from the gut into the bloodstream is usually complete in 30–60 minutes. The body turns alcohol first to acetaldehyde, then to acetic acid and finally to carbon dioxide, providing 7 calories of energy for every gram of alcohol. Slimmers should take this into account when dieting, as well as the tendency of moderate amounts of alcohol to stimulate appetite. Of all the products of the breakdown process, acetaldehyde contributes most to the effects of a hangover, as well as to so-called 'drunken breath'. But the impact can be much more profound. Aldehydes, the family of chemicals to which acetaldehyde belongs, are linked to hot flushes, increased heart rate and palpitations, and can also poison cells and even play a role in the development of cancers. That's why their elimination is crucial to the health of living cells.

The first hint of the importance of genes in our response to alcohol came from studies of drunken mice. Some strains doze off for hours, while others soon come round. Others gaze vacantly at the walls of their cage, while a few rush madly around as if they have seen Marley's Ghost (or the rodent equivalent). The ability of the human body to cope with the spirit of Christmas depends in part on the action of two enzymes, controlled by genes. One, called alcohol dehydrogenase, breaks down the alcohol in the liver; the other attacks acetaldehyde, the even more noxious product that results from the action of the first enzyme. Those unlucky enough to inherit genes that code for a powerful version of the first enzyme and a weak form of the second will suffer immediate and unpleasant side effects after drinking alcohol. Such people are called 'fast flushers'; their faces immediately go red, and they sweat and feel ill whenever they drink, This combination of genes is much rarer in the West than elsewhere in the world, which may be why alcohol plays a uniquely large part in European culture.

Differences in alcohol tolerance are not only due to the body's ability to break down alcohol in the liver. In 1997 a Japanese team discovered a brain chemical that helps animals recover from the effects of alcohol. Without it, they become more drunk and take longer to sober up. The researchers dosed laboratory mice with alcohol, laid them on their backs in troughs and noted how long it took before they managed to stagger to their feet. Mice lacking a crucial enzyme, known as Fyn tyrosine kinase, took twice as long to get up as those with a functioning gene that produced the chemical. Fyn appeared to act in the hippocampus of the brain within five minutes of alcohol arriving there. (Alcohol normally causes brain cells in the region to become less responsive and less likely to fire signals.) 'The absence of the Fyn enzyme makes people more deeply drunk and delays them becoming sober,' says Hiroaki Niki of the Riken Brain Science Institute, who led the research. Individual differences in enzyme levels may explain why some people can hold their drink during a Christmas party, while others reel under its influence.

One of the most disturbing feelings experienced by people who drink to excess is the 'whirling-pit' syndrome. Having had not one but several Christmas drinks too many, and too quickly, sufferers will feel the room spin as they collapse into an armchair or onto a bed. This sensation is caused by alcohol disrupting a balance sensor in the inner ear. The sensor consists of a sac and three semi-circular canals containing fluid. As we move, the corresponding fluid motion is detected by tiny hairs lining the organ, which translate the movement into electrical impulses used by the brain to help calculate balance. To do this, the brain assumes a certain density of the fluid. But as alcohol diffuses into the fluid, this density breaks down, causing an apparent loss of balance, even if there is no actual motion.

This also partly explains why 'hair-of-the-dog' hangover remedies actually work. It is not the absolute density of the balancing fluid that causes disorientation, but changes in its density. During the period of sobering up, when alcohol diffuses out of the fluid, the density changes may be too rapid for comfort, producing a distinctly unsteady feeling. A little alcohol may reduce the rate of change, restoring balance and making you feel less fragile, though there is more to dealing with hangovers than this.

The thirst caused by a hangover may be due to alcohol's diuretic effect: it inhibits a hormone called vasopressin, which controls how the kidneys reabsorb water. As a result, more water passes to the bladder so that drinkers visit the lavatory more often. If you drink around two glasses of wine you lose about twice the equivalent amount of water from the body in the next two hours. Scientists now think they also know why throbbing headaches sometimes result from over-enthusiastic drinking. Headaches do not start in the brain, which has no pain receptors, but in the major blood vessels of the meninges, the membranes between the brain and the skull. For a long time it was thought that the pain was caused by the swelling of these blood vessels. This idea was overturned by techniques that enabled doctors to examine the vessels in conscious patients.

Andrew Strassman and colleagues from the Beth Israel Deaconess Medical Center and Harvard Medical School in Boston have provided direct evidence, based on studies of rats, to suggest that nerve endings in the blood vessels become highly sensitive as a result of being exposed to chemicals in the bloodstream. The same kind of sensitization also primes the nerves to respond to movement, suggesting why, with some hangover headaches, the pain is worsened by coughing or sudden head movement. The chemical culprits for this type of sensitization are usually linked to inflammation and injury. In the case of a hangover, however, they are due to the breakdown products of drink, notably constituents called congeners that are formed during fermentation. Brandy, cheap red rum and red plonk produce the worst hangovers, being high in congeners while relatively low in pure alcohol. Gin and vodka are less likely to produce hangovers for the opposite reasons. A link has been found between the breakdown in the body of one congener, methanol, into formaldehyde and formic acid, and the onset of hangover symptoms. Interestingly, ethanol, or alcohol, blocked this breakdown, offering another explanation of why the hair of the dog may work.

HOW TO COPE

Of course, the best hangover remedy is not to drink in the first place. If you must, then eat. A full stomach retards the passage of alcohol into the small intestine, delaying the effect. That is why a cocktail on an empty stomach has a bigger kick than a couple of drinks after a heavy meal. A glass or two of milk can also help. Another suggested remedy is to take N-acetyl-cysteine, an amino acid supplement available in many health-food stores, which helps replenish the body with glutathione, part of its detox machinery. Exercise, however, doesn't do much to help a hangover: the rate of breakdown of alcohol stays quite steady, at around 21g (¾oz) of whisky per hour for a man weighing 75kg (11¾ stone). And coffee merely acts as a stimulant to counteract some of the alcohol's depressive effects. It is also thought unlikely that traditional remedies, such as sweating in a sauna or taking cold showers, can really help. Sobering up is simply a matter of time.

Sheryl Smith of the Allegheny University of the Health Sciences, Philadelphia, has linked hangovers to changes in a receptor in the brain, in which a messenger chemical called GABA acts to dull nerve circuits. The changes are caused by the increased production of a protein that forms part of the receptor, called the alpha-4 subunit. When this occurs, the GABA receptor is less efficient, so it cannot calm those sizzling circuits quite so easily. A drug that blocks the production of the alpha-4 protein could moderate the symptoms of hangovers. However, a completely effective hangover treatment will probably always be beyond the reach of medical science, according to Ian Calder, a consultant anaesthetist at the National Hospital for Neurology and Neurosurgery, London: the hangover is too complex a condition to cure.

In an editorial for the *British Medical Journal*, Calder maintained that ethanol (alcohol) was only partly responsible for the all-too-familiar hangover symptoms of thirst, headache, fatigue, nausea, sweating, tremor and anxiety. Other factors – such as lack of sleep, smoking, overeating, snoring and any unusual physical or emotional incidents that occur during a drunken night out – must also play a part, making a truly effective treatment for hangovers unattainable. Moreover, he believes that such a remedy is 'arguably undesirable', since the fear of a hangover prompts most people to moderate their alcohol intake. Even moderate amounts can be damaging, so a painful penalty the morning after is in our best interests.

WHAT HAPPENED?

Many people find a heavy drinking session difficult to remember afterwards. If you think you enjoyed a Christmas party but the details are rather hazy, you may be suffering from an imbalance in levels of the protein c-Fos, which provides a general measure of the activity of brain cells. In 1997, scientists from the Scripps Research Institute in La Jolla, California, traced the levels of c-Fos in rats that had consumed the equivalent of three or four drinks in humans. The team found that the protein was stimulated in several regions of the brain, particularly those involved in regulating emotions and behavioural motivation, and in processing sensory stimulation. It was selectively decreased only in the hippocampus, the part of the brain responsible for the formation of complex memories.

This moderate dose of alcohol not only decreased the levels of c-Fos protein in the hippocampus of otherwise-untreated rats, it also blocked the increased activity that typically occurs when the animal is exposed to a new environment. This may explain alcohol's effects in suppressing the ability to remember novel information. A lower dose of alcohol (corresponding to one to two drinks) also decreased levels of c-Fos, but was unable to block the response to a change of environment. This suggests that a lower dose is not enough to disrupt the processing of new information in the hippocampus. These studies help to explain why, after recovering from a binge, one may not remember dancing on a table, or much about the place where drinking occurred. However, due to the way the function of other brain structures was spared, you may still have a feeling that a good (or bad) time was had.

THE ALCOHOL DEVOTEE

Despite these ill effects, alcohol, in small quantities, is good for you. But boost the daily dose of this drug beyond 4 units for a man and 3 for a woman and these health benefits decline. The relationship between death rate and alcohol intake has been the subject of various studies, including a thirteen-year survey of 12,000 male British doctors published in the *British Medical Journal*. The lowest death rates were found among doctors who drank up to 21 units per week, the equivalent of 1½ pints of beer a day.

Alcohol in moderation is thought to boost levels of high-density lipoproteins, or 'good' cholesterol, in the blood, which are associated with a reduced risk of cardiovascular disease. Alcohol may also prevent clots from forming in arteries. One major study by Patrick McElduff and Annette Dobson of the University of Newcastle, New South Wales, published in 1997 as part of a World Health Organization project, concluded from an investigation of around 12,000 heart attacks (one-quarter of which were fatal) that the incidence was lowest among people who consume between one and two drinks daily on five or six days a week.

While such studies have indicated that moderate drinking reduces the likelihood of suffering a heart attack in the first place, there may even be potential benefits to drinking *after* a heart attack. Reporting in the *Proceedings of the National Academy of Sciences*, Vincent Figueredo and colleagues at the University of California, San Francisco, claimed to have found a mechanism by which moderate drinking may reduce the damage caused by a heart attack, thereby improving the chances of survival.

The type of alcohol consumed may also be a factor in determining its beneficial effects. Researchers at the University of New York at Buffalo observed that people who drink wine appear to experience less alcohol-related oxidative stress – cell damage caused by highly reactive chemical intermediaries called free radicals, which play a role in many chronic diseases – when compared with people who drink beer or spirits. 'The difference was small but significant,' their report notes.

Geoff Lowe of the University of Hull believes that psychological factors, notably laughter, also contribute to the benefits of alcohol. By hanging around in local pubs he observed what every drinker knows – that a pint or two makes people jollier. In one survey of 332 social drinkers, humour and laughter were found to be most prevalent in the everyday lives of heavier drinkers. A second experiment showed that people drinking alcohol laughed more than those consuming non-alcoholic drinks while watching a comedy film, *Naked Gun*. In a third study, observers monitored how much young people laughed in bars and pubs. Laughter was again significantly higher for groups of alcohol drinkers. Geoff Lowe concludes: 'Taken together with evidence that laughter can serve as a stress moderator and enhance immune function, our observations suggest that the positive influence of moderate alcohol consumption on health and longevity may at least partly be due to its mood-enhancing and stress-buffering properties.'

Alcohol also acts as a ready-made source of energy, with at least 70 per cent of the calories it provides immediately available for use. A single shot of gin or vodka contains about 125 calories, and a litre of Guinness around 370 calories, for instance. Intriguingly, Colorado State University researchers found no link between moderate red wine consumption – defined as less than 5 per cent of an individual's daily calories – and weight gain. A possible reason may be that calories from alcohol are metabolized differently from other food calories.

But what does all of the above mean in terms of the effects of alcohol on life expectancy? A 1991 study published in the journal *Circulation* calculated that if all heart disease were eliminated, the average life expectancy of a 35-year-old would increase by about three years. If moderate alcohol consumption confers a net health benefit, then some portion of these three added years of life, perhaps a few months, might be attributed to alcohol in moderation. But that is a statistical average, and every individual is different. The same would not apply to younger drinkers, among whom alcohol is thought to cause more deaths due to increased risk of accidents and violent attack than it prevents through reduced risk of heart disease.

As we have already seen, there is also an array of genetic factors that affect our sensitivity to drinking. One day it may be possible to study how each individual responds to alcohol. DNA analysis chips are already on the market that can screen an individual's genetic makeup. Until now, these have been used mainly to detect the risk of inherited disease, but in a few years' time, they will be the ultimate stocking fillers, allowing you to check your own genetic makeup – and assess how you cope with your favourite poison.

But until that time, we all have to take into account the downside of consuming alcohol. First, the calories it contains are 'nutritionally empty', and at high levels of consumption, alcohol can inhibit the absorption of fat and vitamins such as thiamine, riboflavin, niacin and folic acid from the intestine. Alcohol also reduces the body's levels of magnesium, calcium and zinc. Chronic alcoholism is often associated with malnutrition. When drinking exceeds 'moderate' – more than about 3 units a day – it starts to increase a person's susceptibility to cardiovascular disease, for which hypertension, or high blood pressure, is a well-known risk factor. Alcohol is second only to obesity in its contribution to hypertension.

Alcohol is a major cause of liver disease, and has also been linked to certain types of stroke, and to cancer of the mouth, throat and liver. A paper published in 1996 in the journal *Nature Medicine* showed that rats injected with cancer cells developed up to ten times more tumours if first given a large dose of alcohol – equivalent to an alcoholic binge – than if they 'abstained'. The researchers, including Shamgar Ben-Eliyahu of Tel Aviv University, concluded that alcohol suppresses the activity of the immune system's natural killer cells, which are known to play a crucial part in ridding the body of tumour cells.

BOTTOMS UP

A great deal of scientific endeavour has focused on the ill effects of alcohol. During the late 1980s, however, David Warburton of Reading University became increasingly concerned that most research had a gloomy agenda and that few studies were being made of the benefits of eating and drinking. He was alarmed, for instance, by reports that parents in the grip of a high-fibre fad were putting toddlers on low-fat diets, depriving them of the cholesterol they needed to develop their brains and sex hormones. Warburton set up ARISE (Associates for Research Into the Science of Enjoyment). Specialists in fields as diverse as psychology, pharmacology and neurochemistry united in a common purpose: to challenge the hectoring of the health-education lobby by demonstrating that many of the pleasures it frowns on during the Christmas festivities – from eating fatty foods to drinking rich red wines – are good for us in moderation. Feeling guilty about overindulgence, they claimed, could even lead to health problems. 'When people are brooding on their guilt, they become more absent-minded and error-prone,' says David Warburton. 'Chronic guilt can induce stress and depression, leading to eating disorders. It can contribute to infection, ulcers, heart problems and even brain damage.'

With enthusiastic backing from food manufacturers, Warburton searched the world for ARISE recruits, finding support among doctors as far afield as California, Dublin and Italy. He also recruited scientists in Germany and Australia to examine the health benefits of moderate alcohol consumption. The conclusion he drew from this research is that pleasure is in itself an antidote to the harmful stresses of modern living. Even an enjoyable film or a favourite piece of music was found to raise levels of antibody known as immunoglobulin A, indicating a strengthening of the immune system. By contrast, the kind of stress associated with guilt had the opposite effect. It made people more prone to disease by lowering production of lymphocytes, the white blood cells that fight infection. Those of us who are tired, depressed or bullied at work are more likely than happier colleagues to go down with colds and flu.

ARISE surveys suggest that something like four people in ten would enjoy many of their basic pleasures far more if they did not feel so guilty about them. 'Medical evidence that pleasure is good for you is a useful riposte to the moralistic self-righteousness of those who believe that there is only one way to live your life – theirs,' says David Warburton. So long as you do not go overboard, the overindulgence that goes with seasonal good cheer may be good for you.

HIGH-TECH WINES

Within a decade or so, biotechnology may also enable us to enjoy more wine and champagne at Christmas, without the risk of a hangover. This is due to advances in the development of genetically engineered grapevines. The competition to produce these biotech wines is intense, with rival teams at work in the United States, France, Israel and Australia. Nigel Scott, Chief Research Scientist of the CSIRO Division of Horticulture in Adelaide, is head of one effort to introduce new genes into wines. For the purposes of pest resistance and preventing spoiling, this is relatively easy. But the development of a grape specially engineered to produce a spicy Christmas *glühwein* is a much more distant prospect. 'Changing flavour is extremely difficult,' Scott says.

At the CSIRO's Adelaide Laboratory, Mark Thomas and Tricia Franks have studied techniques to introduce new genes into sultana grapevines, paying particular attention to the plant hormones that control regeneration from a few cells to embryo to grapevine. For reasons that are not understood, the parts of the plant that respond best to genetic engineering are filaments taken from the stalk that ends in the plant's anther, the lobes in which pollen matures. Using *Agrobacterium tumefaciens*, a bacterium that naturally transmits genes into plant roots, a marker gene called *Gus* was introduced into cells from the sultana filaments. The engineered cells were cultured in the laboratory into embryos that could be grown into full-size plants. Now the team wants to make sultanas last longer by switching off the gene responsible for the enzyme polyphenoloxidase, which causes them to go brown when they are dried. Nigel Scott is confident that they will soon be able to produce superb, golden-coloured sultanas.

The hunt is also on for other genes to alter. Simon Robinson and Ian Dry, who found the polyphenoloxidase gene, are now searching for grape genes that aid defence against fungal diseases. The genes involved in the ripening process are also being studied to develop grapes that give improved berry softening, sugar accumulation and synthesis of flavour compounds. The team also hopes to extend the work to Chardonnay, Pinot and Shiraz grapes.

One aim close to the heart of the Christmas tippler is to produce a grape with reduced sugar, and thus reduced alcohol, content. The wine would be superior in terms of taste to low-alcohol wines now on the market. Then Santa can have two bottles of wine with Christmas lunch and still be able to drive the sleigh in a straight line on the way home.

ten

Christmas Blues and Seasonal Moods

He was checked in his transports by the churches ringing out the lustiest peals he had ever heard. Clash, clang, hammer; ding, dong, bell. Bell, dong, ding; hammer, clang, clash! Oh glorious, glorious!

Running to the window, he opened it and put out his head. No fog, no mist; clear, bright jovial, stirring, cold; cold, piping for the blood to dance to; Golden sunlight; Heavenly sky; sweet fresh air; merry bells. Oh, glorious! Glorious! ... Christmas Day!

Charles Dickens, *A Christmas Carol*

Smiling, happy people are part of Christmas mythology, reflecting a belief that human behaviour is somehow different at this time of the year. The same belief is found throughout the work of Charles Dickens. From his first account of Christmas festivities in *Sketches by Boz* to the beatification of Scrooge in *A Christmas Carol*, Dickens emphasizes again and again that Christmas is a time for social altruism, warmth and friendship. In our own day, scientists have produced quantifiable evidence that people are affected by the season and the celebrations themselves, beyond obvious manifestations of Christmas spirit such as hangovers, family gatherings and expanding waistlines. Some of these influences are positive, others negative and a few are downright odd.

CHRISTMAS GLOOM

Holidays and festivals are widely believed to increase the incidence and expression of psychiatric disorders. The Swiss even have a name for it, *Weihnachtscholer* or 'Christmas unhappiness'. It's one of the great yuletide ironies: that in a season celebrating joy, fellowship and charity, the 'holiday blues' drive the suicide rate upwards. Although recent research has cast doubt on this idea, the suspicion still remains that, for some, Christmas can be the season of bad cheer and misery. Psychologists have speculated on whether Christmas analogies, symbols and fantasies somehow bring on mental illness. Perhaps even the slackening of the prohibitions against self-indulgence at Christmas can trigger depression. Or maybe the seeds of seasonal misery are sown when hopes raised by magical wish-fulfilment are dashed, after the holiday, by the frustrations of everyday life. Feelings of loneliness and inadequacy may also be emphasized when a family Christmas falls short of expectations.

Over the years, some barmy psychoanalytic explanations have been put forward for these gloomy yuletide side effects. One, outlined in the summer of 1955 by psychiatrist Bryce Boyer in the *Journal of the American Psychoanalytic Association*, put the depression down to 'reawakened conflicts related to unresolved sibling rivalries ... it is suggested tentatively that partly because the holiday celebrates the birth of a Child so favoured that competition with Him is futile, earlier memories, especially of oral frustration, are rekindled.' Boyer based his ideas on a sample of seventeen patients, including 'Mrs W', a thirty-year-old childless housewife and 'inactive Episcopalian', who had become depressed and preoccupied with religious thoughts by mid-December. She had always been convinced that she was unwanted and had blamed this on being female. According to Boyer, her ruminations centred on one theme: 'If only I would turn to religion, God would give me a penis.'

Bryce Boyer also cited the case of 'Miss X', a 34-year-old Episcopalian virgin, newspaper editor and lesbian, who had also become depressed, thinking of her minister as God and considering herself to be 'Christ's rival for the position of His favourite child'. Boyer's verdict: 'This woman's unconscious aim had been to acquire a penis with which she could satisfy her mother sexually and obtain in return the permanent role of mother's favourite baby, with constant access to a full breast.' Then there was the case of 'Mrs Y', a 32-year-old Jewish mother who became depressed around Christmas and very much wanted a Christmas tree. Boyer reasoned that she 'equated tree and penis'. He concluded that his patients 'uniformly' sought to obtain penises 'with which they imagined they could woo their mothers to give them the love which they felt had been previously showered upon their siblings ... They at times identified with Christ in an attempt to deny their own inferiority and to obtain the favouritism which would be His just due.'

Equally bizarre was the attempt in 1944 by the psychologist Richard Sterba to compare behaviour around Christmas to the customary activities surrounding childbirth: the preparatory excitement, the secret anticipation, the last-minute flurry of preparation, the prohibition about entering the rooms containing the gifts, and the relief of tension afforded by the delivery of the gift, whether that of a baby or a sweater knitted by an aunt. 'It is not surprising,' Sterba wrote, 'that the presents come down the chimney, since fireplace and chimney signify vulva and vagina in the unconscious and the child-present thus comes out of the birth canal. This casts some light on the figure of Santa Claus. He, no doubt, is a father representative ...' He concludes that Christmas stirs up unconscious fantasies and unresolved conflicts about childbirth, causing susceptible people to develop mental illness around this time.

Another article, presented in 1954 to the American Psychoanalytic Association in St Louis, put forward the idea of a holiday syndrome, lasting from Thanksgiving until after New Year's Day, characterized by depression, 'diffuse anxiety', nostalgia and 'wishes for magical resolution of problems'. This syndrome was said to reach a zenith at Christmas, probably because sufferers had difficulties establishing close emotional ties and, as a result, felt isolated, lonely and bored. This at least seems more reasonable than Boyer's Christmas neurosis.

Today, from the beginning of December, Christmas blues regularly feature in the media, with acres of newsprint expended on how to deal with them. But does this coverage reflect an actual peak in suicides? No, according to several research studies. In Britain, one study of 22,169 attempted suicides over nineteen years revealed that the rate among women dropped to three-quarters of the usual level during the Christmas period. There was no evidence of a monthly or seasonal variation among men.

Similar results from the United States confirm that the 'syndrome', if it exists, does not produce a strong enough effect to show up in monthly statistics on suicide. Studies carried out in the sixties and seventies also found that the numbers of suicides in December and January were average or low. One investigation conducted in North Carolina was published in the *Archives of General Psychiatry* in 1974. William Zung and Richard Green studied 3,672 suicides and 3,258 admissions to the Durham Veterans' Administration Hospital Psychiatry Service. They found no statistically significant increase in psychiatric admissions to hospital or in suicides at Christmas, or during any other holiday period. Overall, American statistics compiled over the past two decades show that December records the fewest suicides of any month. Suicide rates tend to rise in spring and summer and edge down in autumn and winter. 'What you find is not an increase but a drop in suicide during the holidays,' says David Phillips, a sociologist who has studied these seasonal trends. In 1991, for example, the months of April and June tied for the highest average daily suicide count, followed in order by August, July, May and September. December ranked last. 'The holidays appear to be providing some psychological and social protection against suicide, but the nature of this protection is currently unknown,' Phillips concludes after analysing the timing of more than 180,000 suicides from the late seventies.

WHY ARE WE SAD?

The statistical decline in suicide rates during the holidays does not necessarily mean that the incidence of depression also declines. It is conceivable that while fewer people actually carry out the threat to kill themselves, the number who come to a hospital emergency room or a doctor's surgery complaining of depression rises during the holidays.

There is undoubtedly a propensity for some people to feel gloomy as a result of seasonal affective disorder (SAD), a term first coined in 1984 by Norman Rosenthal, a psychiatrist with the US National Institute of Mental Health. The syndrome is characterized by recurrent depressions that occur annually as a result of the decline in daylight hours, which triggers a change in brain chemistry among many people. For a significant proportion of the population living at high latitudes, winter days can mean misery. Winter after winter, these people experience lethargy and fatigue, sadness and despair. According to Rosenthal, SAD disrupts personal relationships, and causes victims to overeat and to become indifferent towards their jobs.

For a long time, the winter blues were written off as psychiatric curiosities. But in the past few years a number of publications in scientific journals have supported Rosenthal's ideas, representing an acknowledgement by the medical community that SAD is a real illness. It is now recognized that women are four times more likely than men to be affected by the disorder, which usually starts after puberty and diminishes after menopause. Rosenthal speculates that female reproductive hormones somehow sensitize the brain to the effects of light deprivation. Incidence of SAD is directly related to latitude, varying according to the day–night cycle and affecting mostly those farthest from the Equator. American studies have shown SAD symptoms in about 10 per cent of people in New Hampshire, which has long winter nights, while only about 1.5 per cent are affected in Florida, where winter days are longer.

Doctors have found that these dark moods can be banished to some extent by exposure to intense light. This may explain, for example, why bright white snow can help lift winter gloom. Light treatment can even be effective in the middle of the day, and the reason it works has puzzled scientists. Recently, however, a milestone experiment on hamsters by a team at Northwestern University, Evanston, Illinois, has offered new insight into how light adjusts brain chemistry to make winter and Christmas more tolerable for SAD sufferers. For a number of years it has been known that a brain-signalling chemical called serotonin is involved in depression. Fred Turek and his colleagues at Northwestern found that pulses of light administered in the middle of the day altered the way hamsters responded to the effects of serotonin on the biological clock. These results indicate that SAD may be caused not by a simple lack of light but by the effect of that light on serotonin metabolism.

VIVA CHRISTMAS

The festive celebrations during the dark days of winter may have one unexpected bonus for those who are near death. There is growing evidence that the moribund can successfully 'bargain with God' or exercise willpower to manage to live until some important occasion, such as a festive meal with close family. David Phillips, working with Elliot King, set out to test whether a dying person can indeed survive for a few days longer than they might otherwise have done in order to join in a seasonal celebration, such as Christmas or Hanukkah. They compared deaths among Jews and non-Jews before and after the Jewish holiday of Passover. The subject was attractive because Passover shifts around the calendar by about four weeks a year, allowing separation of holiday effects from purely seasonal effects, particularly changes in the hours of available daylight. The most important event of Passover is a ritual dinner called the Seder, which celebrates the Exodus of the Israelites from Egypt, their receiving of the Torah and their entry into the land of Canaan. More than three-quarters of American Jews attend the Seder.

To form a clear picture, Phillips and King needed large numbers of subjects to take part. California death certificates do not list the faith of the deceased, but they focused on 1,919 death certificates of men who had characteristically Jewish names, such as Cohen, during the weeks before and after Passover from 1966 to 1984. For one set of controls they used a list of surnames, such as Rose and Green, that are common in both Jewish and non-Jewish populations; for other controls, the researchers isolated a group of people coded as Japanese or Chinese on death certificates. They discovered a dip in deaths in the week before Passover among the Jewish group and an 8 per cent increase in the week after. Then, among 625 men with 'unambiguously' Jewish names, they found a 25 per cent increase in the week after Passover. When it came to weekend Passovers, when family gatherings are likely to be larger, there were 61 per cent more deaths in the week after than in the week before. They found no such differences among the Oriental control group.

This finding, which the authors called the 'Passover effect', applied solely to adult men and was absent in adult women and in children under the age of four. The effect appeared in deaths from all three main causes (heart disease, malignant tumours and cerebrovascular disease), but not in deaths from other natural causes or from external causes. Why, then, did the Angel of Death 'pass over' the Jewish households until after the holiday? This, Phillips and King argued, is 'consistent with two hypotheses: that the "will to live" is associated with reduced mortality, and that communal social events can have a beneficial impact on the course of a disease.'

This is by no means a modern phenomenon. Phillips and King have pointed out that both John Adams, the second US president (born in 1735), and the third president, Thomas Jefferson (born in 1743) managed to live until 4 July 1826, the fiftieth anniversary of the signing of the Declaration of Independence. According to Jefferson's physician, his last words were, 'Is it the Fourth?'

However, some experts were unconvinced that the answer to these seasonal changes in death rate lies in the individual's urge to live to see the big day. Michael Baum of King's College School of Medicine, London, commented in *The Lancet* (1988) that diet offered an alternative explanation of the statistics. 'The Passover diet among observant Jews has to be free of all leavening and thus is highly costive,' he wrote. A diet rich in matzo (unleavened bread) contains little bran and has a constipating effect which results in straining at stool, raised blood pressure and increased risk of heart attack or stroke. Baum also raised another issue concerning women. Deaths of Jewish women are *not* affected by the Passover, and the Californian team argued that this is because they take a lesser part in the ceremony. But Baum countered that the Passover drives Jewish women near to breakdown: 'Large numbers of wives from the ultra-orthodox community crack up under the psychological strain of preparation for the Passover, seeking refuge in the hospital ward until the Angel of Death passed over the household and normal services could be resumed.'

Despite this, Phillips and King have found still more evidence in support of their theories. Every year for the past twenty-five years, in the week immediately before the traditional Chinese harvest-moon festival, they have observed evidence of the same phenomenon in their local Chinese community. They described their findings in 1990 in the *Journal of the American Medical Association*. The death rate for older women, who play a central role in the celebration of the holiday, suddenly dips by as much as 35 per cent, as if those who would have died postpone their passing to be part of the occasion. Then, after a corresponding increase in the death rate immediately after the festival, the rate returns to its usual level. These observations seemed to confirm the earlier findings among Jewish men before and after Passover. A control group, meanwhile, drawn from people for whom the Chinese holiday had no symbolic significance, showed no such dip.

This phenomenon cannot be explained by conventional hypotheses, says David Phillips. They do not account for the drop in the number of deaths before the festival. The Chinese women are not dying in greater numbers after the holiday because of extra stress or overeating. Nor can the behaviour of either the Chinese women or the Jewish men be explained by any fixed monthly pattern in mortality, because both holidays move around the calendar from year to year. The 'causal pathway' linking psychological and biological events may be far broader than previously recognized. Personal events of great symbolic importance – a fiftieth wedding anniversary, for example – may also have demonstrable short-term effects on mortality. 'The best available explanation is that deaths of some people are postponed until they have reached an occasion that is important to them,' says Phillips. The moribund put off death for these 'symbolically meaningful occasions', which include Christmas.

SEASONAL JOLLINESS IS GOOD FOR YOU

Let's assume for a moment that people are jollier than usual and celebrate more during the Christmas season. Are there any health implications? Studies at the Institute of HeartMath in Boulder Creek, California, published in 1995 in the *American Journal of Cardiology*, suggest that while anger can put the heart at risk, 'Sustained positive feelings of the kind (presumably) in good supply over the holiday period can be used to protect against heart attacks and high blood pressure.' A relaxation technique designed to manage mental and emotional stress has been shown to affect the autonomic nervous system, which controls heart rate and breathing. Alan Watkins, a doctor at Southampton General Hospital, believes that 'Christmas spirit' may have a similar, if less marked, beneficial effect.

Although I admit this is stretching things somewhat, some studies on mice seem to suggest that an enjoyable experience, such as Christmas festivities, can also enhance brain function. A number of studies over the past four decades have shown that placing laboratory animals in a stimulating environment boosts performance in standard tests, such as the ability to negotiate mazes. However, the biological mechanisms underlying these improvements have been hotly debated. In 1997 a team at the Salk Institute in La Jolla, California, led by Fred Gage conducted a study to investigate the effects of environment on the brain. Gage's colleagues Gerd Kempermann and Georg Kuhn separated 21-day-old mice into two groups: one housed in 'standard' laboratory conditions, the other in a large cage 'enriched' with tunnels, toys, an exercise wheel and food treats such as apples, popcorn and 'whole-grain nibble bars'. (This is the equivalent of comparing a group of people who have a boring time over Christmas with another who enjoy the celebrations.)

After forty days, the two groups were compared in a water-maze test. As expected, the group raised in the 'mouse-Christmas' conditions performed significantly better than the control group. The scientists also found that the hippocampus, the brain region linked to memory, contained an average of 40,000 more nerve cells in the 'enriched' mice than in the controls. 'We were overwhelmed by the magnitude of the increase, which represents a gain of 15 per cent in the number of these nerve cells,' Gage says. He thinks that the mice in the enriched group are not necessarily *generating* more brain cells, since brain cells seem to be dividing at the same rate in both groups. Instead, an enriched environment appears to foster the survival of new brain cells.

Several days before the end of the experiment, when the

mice were around sixty days old, some of them were injected with a dye that would make new brain cells fluorescent and thus identifiable under the microscope. In the enriched animals, the survival rate of newborn cells was 60 per cent higher than in the control group. The results are remarkable because the experiments were not carried out in infant mice, whose brains one would expect to be formable, but rather in older animals. If the human brain works like that of a mouse, adult humans can also improve the 'architecture' of their brains by enriching their lifestyle with Christmas celebrations.

CHRISTMAS EXCITEMENT

But why celebrate at all at this time of year? One explanation dates back to prehistoric fascination with the Sun. Another may have been uncovered by smell researchers, who believe that chemicals in human sweat exert an effect on the brain over and above that of religion, excessive consumption of food and drink, and yuletide bonhomie. The chemicals are called semiochemicals or pheromones. Scientists have provided convincing evidence that these airborne chemical signals can affect the human body, with the discovery in 1998 by researchers at the University of Chicago that armpit secretions in women cause menstrual changes in other women. The physical and mental reactions triggered in us when we smell sweat are still not fully understood, but they may have a bearing on seasonal festivities.

A study reported in 1987 by scientists at the Monell Chemical Senses Center in Philadelphia suggested that every December sees a peak in the levels of two chemicals found in male sweat, androsterone and androstenol. 'The peak is right at the end of December,' comments David Kelly of the Department of Chemistry at the University of Wales, Cardiff, who believes that these sweat chemicals may indeed be linked to Christmas spirit. They are not responsible for the particular smell of sweat but nonetheless are musky-smelling steroid mol-

ecules. They are also found in male pigs, in special white gluti-
nous saliva, and play a role in porcine fun. The boar smears the
saliva on the face of the sow to determine if she has ovulated
recently and is ready to mate. Androstenone, related to
androstenol, can even be bought in spray form by pig farmers
who need an aphrodisiac to put their charges in the right
mood.

Androstenone, a sister substance, does not look particularly
promising as a human sex attractant, since some find its smell
off-putting. However, when one of several chairs in a doctor's
waiting room was sprayed with it, women preferred to sit on
that chair. Another experiment, conducted two decades ago,
showed a similar effect, when the steroid was sprayed on seats
in a theatre.

Of the female anatomy, breasts exude the most androstenone
and one suggested explanation for its mood-altering effect is
that its aroma is linked to memories of being breast-fed and
thus to mother-baby bonding. 'I have speculated that the cause
of Christmas spirit may be a sudden increase in production of
this material,' says David Kelly. Because December has seen
feasting since ancient times, Kelly argues that production of
these steroids may be one reason why many people feel more
excited at this time of year.

The beneficial effects of Christmas don't stop there. Those
who are born on Christmas Day may well have a better start in
life than other people, thanks to a phenomenon called BIRG.
The initials stand for 'basking in reflected glory', a term coined
by psychologists to denote the human tendency to associate
with success.

The most influential scientist of all time, Sir Isaac Newton,
born on Christmas Day in 1642, may have been a product of
the BIRG effect. A team from the University of California,
Davis, and the Israel Institute of Technology in Haifa has investigated this relationship between the dates of national holidays,
particularly Christmas, and the birth dates of successful people.

Among those born during the weeks centring on national holidays, the California scientists discovered that a disproportionate number of celebrities were born on the holiday itself. This phenomenon was more marked when it came to Christmas than on either 4 July or 1 January, and the findings were true both for entrants in *Who's Who* and for members of the US Congress.

According to the BIRG theory, the association of the birthday and 'positively evaluated stimuli' – a national holiday – enhances that person's image. Perhaps the parents of a child born on Christmas Day believe that he or she is special, like baby Jesus, and will have an edge in life as a result. Perhaps, once the bond between holiday and birthday is forged, it boosts confidence, so that the feeling that this person will grow into someone special becomes a self-fulfilling prophecy. Albert Harrison, Nancy Struthers and Michael Moore, who conducted the study, decided to take this research one step further and study whether people born around Christmas went on to become religious high-flyers. Although the findings were tentative, they did support the BIRG theory: in comparison with low-ranking Christian clergy, high-ranking clergy were more likely to have been born on 25 December.

THE FAITH FACTOR

Today, there seems to be quantifiable evidence that those who respect religious traditions, and thus take the seasonal festivities seriously, can expect to lead a healthier life. In 1872, Francis Galton, the father of human genetics, wrote in his *Statistical Inquiries into the Efficacy of Prayer* that he could find no evidence that prayer is effective. By that he meant he found no scientific grounds for believing that prayers are answered, yet he conceded that prayer can strengthen resolve and relieve distress. A wide-ranging survey of scientific evidence for the importance of the 'faith factor' in fighting disease has been conducted by

Dale Matthews of Georgetown University School of Medicine in Washington DC. He reviewed 212 studies and found that in three-quarters of these religion was shown to have a positive effect, notably in cases of substance abuse, alcoholism and mental illness. Religious people appeared to enjoy a better quality of life than nonbelievers, to be less prone to illness, and if they did become ill, to have better chances of survival. Seven out of ten studies of serious illness showed that religious people lived longer. 'Whether one looks at cancer, hypertension, heart disease, or physical functioning, most of the studies demonstrate a positive effect,' Matthews says (see Appendix).

Studies of the faith factor do face profound methodological difficulties. To some, 'religion' means personal belief. To others, it means church attendance or conforming to traditional rituals. The added problems of disentangling the interplay between religion and mental health, the effects of religion on lifestyle, and how a decline in physical health may make someone turn to religion as a durable source of hope, meaning, and purpose, mean that this field is fraught with difficulties for the researcher.

Nonetheless, a number of major studies consistently point the same way. One study of 4,000 randomly selected people in North Carolina suggests that older people who attend religious services are less depressed and in better physical health than those who do not. The study was funded by the US National Institute on Ageing, the largest of its kind, and its lead author Harold Koenig concluded that 'Church-related activity may prevent illness both by a direct effect, using prayer or scripture reading as coping behaviours, as well as by an indirect effect through its influence on health behaviours.' He has gone on to outline his ideas in the book *Is Religion Good for Your Health?*

Koenig first became interested in the faith factor through anecdotal evidence. 'I was seeing patient after patient using religion in some way to help them cope,' he says citing activities like praying, saying the rosary and reading the Bible. 'I got intrigued by whether it was really helping them.' In nearly a dozen studies, in locations ranging from prisons to hospitals to individuals' own homes, he discovered that it was.

Some aspects of the influence of religion on health are obvious. As Koenig has found in his own studies, praying and other coping strategies can help older people moderate harmful stress. Religious people may also be more compliant and thus more likely to adhere to regimes prescribed by their doctors. Older church-goers generally have an entire congregation watching out for them. Koenig also believes that the devout are less likely to take health risks: 'Active religious participation may indirectly prevent health problems due to poor diet, substance abuse, smoking, self-destructive behaviours, or unsafe sexual practices, because these activities are discouraged by most religious groups.'

Many scientists suspect that there is a connection between state of mind and the health of the body. However, even if this is accepted, a 'chicken-or-egg' problem arises: does an optimistic religious outlook promote a healthy body or vice versa? A study of 1,000 people undertaken to investigate the relationship between religion and mental health, and published in the *American Journal of Psychiatry*, showed that the degree of religious belief does correlate with *subsequent* mental health. Moreover, the stronger a patient's religious faith, the faster he or she recovered from depression, especially if the patient was disabled or chronically ill. In a 1998 study of eighty-seven depressed people hospitalized for medical conditions such as heart disease and stroke, those who scored high in 'intrinsic religiosity', as measured by a scientifically validated questionnaire, recovered faster than other patients. 'This is the first study to show that religious faith by itself, independent of medical intervention and quality of life issues, can help older people recover from a serious mental disorder,' says Harold Koenig. 'What we may be able to do some day is to predict – from the strength of belief and change in symptoms – who is going to get better and who is going to get worse in terms of depression and so on.' The salutary effect of religion isn't limited to mental health, Koenig adds. Preliminary data suggest that, in old age, religious people have lower blood pressure and lower death rates from coronary heart disease.

You do not have to invoke divine intervention to explain this. Because religious beliefs and behaviour may help a person cope with stress, the negative effects of stress on the body may be reduced or avoided. Depression, anxiety or psychological turmoil result in the release by the adrenal glands of the stress hormones cortisol and epinephrine, chemicals which prepare the body either to confront or escape the danger (the so-called 'fight-or-flight' response). When such stress is prolonged over weeks or months, the release of these substances can adversely affect the body's immune and cardiovascular systems, increasing the risk of disease. Harold Koenig continues: 'Some of the first evidence of a connection between religious involvement and immune-system function has recently been discovered in a sample of 1,700 subjects in a study at Duke University Medical Center, where low church attendance has been associated with higher levels of interleukin-6, an inflammatory cytokine indicative of immune system disregulation.'

People with colds who have more extensive social ties produce less catarrh, are able to clear their nasal passages more effectively and shed fewer cold viruses than people with fewer social ties, according to a study published in 1997 in the *Journal of the American Medical Association* by Sheldon Cohen of Carnegie Mellon University, Pittsburgh. He and his colleagues exposed 276 healthy volunteers, aged between eighteen and fifty-five, to cold viruses to examine this association. They cite information from other studies showing that people with multiple ties to friends, family, work and community live longer than those who are members of fewer social groups. Their findings would seem to suggest, contrary to expectations, that the more people you meet at Christmas, whether in church or at a party, the less likely you are to get a cold. The authors theorize that participation in a more diverse social network may increase the motivation to care for oneself by promoting feelings of self-worth, responsibility, control and meaning in life. Greater diversity of social contacts has also been related to decreased incidence of anxiety, depression and 'nonspecific psychological distress'.

In a similar way, religious belief can help cope with the feeling of hopelessness that has been linked to hardened arteries by Susan Everson of the Public Health Institute, Berkeley, California. Her four-year study of 942 middle-aged men, reported in 1997 in the journal *Arteriosclerosis, Thrombosis and Vascular Biology*, links this state of mind – defined as feeling a failure or feeling that the future is uncertain – to a faster progression of atherosclerosis, a progressive disease in which fat, cholesterol, cellular waste products and calcium collect in the blood vessels, impairing their ability to deliver oxygen and nutrients to the body and setting the stage for a heart attack or stroke. Everson says that among those who reported high levels of hopelessness, there was a 20-per-cent greater increase in atherosclerosis than in those with a brighter outlook. 'This is the same magnitude of increased risk that one sees in comparing a pack-a-day smoker to a nonsmoker,' she comments.

These findings in turn pose another question. How far would the health benefits of religious belief extend to other activities in which there is a lot of social networking or some kind of lifestyle philosophy, for instance, yoga or transcendental meditation classes? 'They have not been compared in any head-to-head manner that I am aware of,' Harold Koenig says. 'My sense is that yoga, TM classes, or social networking – in the absence of a belief system that stresses life commitment and accountability – simply will not work as well.'

If the health benefits of religion were satisfactorily proved, this would raise other issues. Would doctors prescribe a faith for medical reasons? And if they did, would it really count (at least in the eyes of religious leaders) if someone believed because they wanted to live longer rather than because they felt the religious truth of a certain doctrine? Would some religions be more effective than others? And would the discovery herald a return to that mythical Christmas of yesteryear, one where faith – not consumerism – held centre stage? Dale Matthews of Georgetown University School of Medicine believes that doctors should encourage patients with 'autonomous' religious beliefs. 'Will we get to the day when there will be a Surgeon General's warning that those who skip worship services are more likely to die? I don't know,' he admits.

Finally, there is the question of whether the scientific perspective undermines the religious beliefs that lie at the heart of Christmas for so many people. The fear that science would turn people away from God was investigated as early as 1916, when an oft-cited survey by James Leuba of Bryn Mawr University, Pennsylvania, found that 60 per cent of American scientists did not believe in God. The study caused a scandal at the time, prompting warnings from politicians about the evils of modernism and accusations that scientists were leading college students away from religion. Leuba himself predicted that disbelief among scientists would increase in the future. However, recent research, repeating his 1916 survey word for word, has proved him wrong. The proportion of scientists who believe in a god has remained almost unchanged in the last eighty years, despite the enormous leaps of discovery made this century. Edward Larson from the University of Georgia and colleague Larry Witham from Burtonsville, Maryland, questioned 600 scientists listed in the 1995 edition of *American Men and Women of Science* and arrived at the same result as Leuba – about 40 per cent of scientists believe in a God. The future of Christmas and Hanukkah in our increasingly technological age seems assured.

eleven

Santa's Science

> I would stay awake all the moonlit, snowlit night to hear the roof-alighting reindeer and see the hollied boot descend through soot. But soon the sand of the snow drifted into my eyes, and, though I stared towards the fireplace and around the flickering room where the black sack-like stocking hung, I was asleep before the chimney trembled and the room was red and white with Christmas. But in the morning, though no snow melted on the bedroom floor, the stocking bulged and brimmed; press it, it squeaked like a mouse-in-a-box; it smelt of tangerine; a furry arm lolled over, like the arm of a kangaroo out of its mother's belly.   Dylan Thomas, *Conversation about Christmas*

On the Christmas cards, it all looks so effortless. Apart from the occasional slip-up with drunken reindeer, narrow chimneys and blizzards, Santa manages to deliver millions of gifts on Christmas Eve, maintaining his smile and benevolence all the while. His support team: a few reindeer and a handful of diligent elves.

Clearly, only an innocent child would swallow this propaganda, a fantasy peddled by generations of Christmas cards to divert attention away from what is undoubtedly the most spectacular research-and-development outfit this planet has ever seen.

I like to think that somewhere under the North Pole there is an army of scientists experimenting with the latest in high-temperature materials, genetic computing technologies and warped geometries of time and space, all united by a single purpose: to make millions of children happy each and every Christmas. Put yourself in Santa's fur boots: how does he know where children live, and what gifts they want? How can he fly in any weather, circle the globe overnight, carry millions of pounds of cargo and make silent, rooftop landings with pinpoint accuracy? Some years ago, *Spy Magazine* examined these issues in an article that has since proliferated across the Internet. The magazine concluded that Santa would require 214,200 reindeer and, with the huge mass of presents, would encounter 'enormous air resistance, heating the reindeer up in the same fashion as a spacecraft re-entering the earth's atmosphere.' In short, it continued, 'They will burst into flame almost instantaneously, creating deafening sonic booms in their wake. The entire reindeer team will be vaporized within 4.26 thousandths of a second. Santa meanwhile, will be subjected to forces 17,500.06 times greater than gravity ... In conclusion – if Santa ever *did* deliver presents on Christmas Eve, he's dead now.'

The point is, Santa is not dead. He delivers presents every Christmas Eve, as reliably as Rudolph's nose is red. If he overcomes the kinds of problems outlined above, it can only be with the aid of out-of-this-world technology.

SANTA'S CHALLENGE

Santa has a huge market: there are 2,106 million children aged under eighteen in the world, according to the United Nations Children's Fund UNICEF. Given the pagan origins of the festival, and the emphasis on charity, we can assume that Santa will deliver presents to each and every child and not just Christian children or the 191 million who live in industrialized countries. It is Christmas, after all.

Assume there are 2.5 children per house. That means Santa has to make 842 million stops on Christmas Eve. Now let's say these homes are spread equally across the land masses of the planet. The Earth's surface area is, given a radius of 6,400km

(3,986 miles), 510,000,000 sq km (196,600,000 sq miles), calcu-lated as radius squared, multiplied by 4π. Only 29 per cent of the surface of the planet is land, so this narrows down the pop-ulated area to 150,000,000 sq km (57,900,000 sq miles). Each household therefore occupies an area of 0.178 sq km (0.069 sq miles). Let's assume that each home occupies the same sized plot, so the distance between each household is the square root of the area, which is 0.42 km (0.26 miles).

Every Christmas Eve, Santa has to travel a distance equiva-lent to the number of chimneys − 842 million − multiplied by this average spacing between households, which works out to be 356 million km (221 million miles). This sounds daunting, particularly given that he must cover this distance in a single night. Fortunately, Santa has more than twenty-four hours in which to deliver the presents. Consider the first point on the planet to go through the International Date Line at midnight on 24 December. From this moment on, Santa can pop down chimneys. If he stays right there, he will have twenty-four hours to deliver presents to everyone along the Date Line. But he can do better than this, by travelling backwards, against the direction of rotation of the Earth. That way he can deliver pre-sents for almost another twenty-four hours to everywhere else on Earth, making forty-eight hours in all, which is 2,880 minutes, or 172,800 seconds.

From this, one can calculate that Santa has a little over two ten-thousandths of a second to get between each of the 842 million households. To cover the total distance of 356 million km (221 million miles) in this time means that his sleigh is moving at an average of 2,060 km (1,279 miles) per second. Ignoring quibbles about air temperature and humidity, the speed of sound is something like 1,200 km (750 miles) per hour, or 0.3 km (0.2 miles) per second, so Santa is achieving speeds of around 6,395 times the speed of sound, or Mach 6395.

When a sleigh, or indeed any object, exceeds the speed of sound, there will be at least one sonic boom. This is a shock wave sent out when the sleigh catches up with pressure waves it generates while moving, explains Nigel Weatherill of the University of Wales, Swansea, who helped the Thrust Supersonic Car to break the sound barrier in 1997.

Santa, however, does not generate any sonic booms on Christmas Eve. In his book *Unweaving the Rainbow*, Richard Dawkins says he has used this fact to disprove the existence of Santa to a six-year-old child. To a biologist this may indeed seem persuasive but, to an aerodynamics engineer, it suggests that Santa has found a way to suppress sonic booms. For example, says Nigel Weatherill, perhaps Santa cancels the peaks and troughs in the shock wave with troughs and peaks of 'anti-sound' generated by a specialized speaker on his sleigh.

The speed of light is absolute and cannot be exceeded (however, see below), so we should check that Santa is not breaking cosmic law. The usual figure for the speed of light is 300 million metres (984 million feet) per second which, given that there are 1,000 metres per kilometre (5,280 feet per mile), works out to be 300,000 km (186,000 miles) per second. Santa is comfortably within this limit, travelling at around one-145th of the speed of light – too slow to worry about the implications of Einstein's theory of relativity. This assumes, however, that Santa throws the presents down each chimney while passing overhead. In fact, he stops at each house so that he has to achieve double the speed calculated above (from a standing start, he has to travel the distance between each house in two-10,000ths of a second). That means going from 0 to 4,116 km (2,558 miles) per second in two-10,000ths of a second, an acceleration of 20.5 million kilometres (12.79 million miles) per second per second, or 20.5 billion metres (67.3 billion feet) per second per second.

The acceleration due to gravity is a mere 9.8 metres (32ft) per second per second, so the acceleration of Santa's sleigh is equivalent to about two billion times that caused by the gravitational tug of the Earth. Given that Santa is excessively overweight, say around 200kg (30 stone), the force he will feel is his mass times his acceleration: around 4,000 billion newtons. Even fighter pilots can't cope with accelerations more than a few times that of gravity: they have to use special breathing and so called g-suits to keep the blood in their head. Santa would have to cope with around 2 billion times this acceleration. As the physics professor Lawrence Krauss put it, that would reduce our fat friend to 'chunky salsa'.

Krauss has considered similar problems in his work on the physics of *Star Trek*. The starship *Enterprise* gets by with devices called 'inertial dampers' to cushion the forces that Captain Kirk feels in the seat of his pants. Santa has to resort to similar tactics, creating an artificial world within his sleigh in which the reaction force that responds to the accelerating force is cancelled, perhaps by some kind of gravitational field.

There is one other problem Santa has to contend with. His cargo of toys. Assuming that each of the 2,106 million children gets nothing more than a medium-sized construction set (900g or 2lb), he has a load of 1,895 million kg (4,212 million lb) to convey. Then there is also his supply of fuel to achieve these huge speeds. Any way you look at it, Santa has some serious hurdles to overcome.

TELEPORTING SANTA

What if Santa could visit homes using a transporter of the kind routinely deployed by Captain Kirk to beam down from the Starship *Enterprise* to the surface of an alien planet? The first edition of this book prompted enquiries along these lines from readers inspired by news that scientists had carried out the first full demonstration of teleportation – a real-world version of *Star Trek* technology dreamed up a few years ago by, among others, Charles Bennett of IBM.

Teleportation exploits quantum theory, the most revolutionary scientific theory of the twentieth century. This theory was developed by European physicists who realized that the previous 'classical' theories of physics did not work when applied to subatomic particles. However, the endless newspaper articles that greeted the first teleportation experiments mostly ignored quantum theory, relying on references to *Star Trek* to grab the interest of their readers. By the end of this section you will understand why. Quantum theory paints a decidedly potty picture of the way the world works.

The subject of the first teleportation experiments in 1998 was limited to a single particle of light – a photon. The team that accomplished the feat, led by Anton Zeilinger at the University of Innsbruck, did not manage to teleport the entire particle but the 'quantum state' of the photon. That is, a vital statistic such as its state of energy or polarization. Although this is a far cry from Captain Kirk, boots and all, Zeilinger's research raises a fascinating possibility: the research and development powerhouse that supports Santa's seasonal activities may have refined teleportation technology to the point that our plump friend can beam down chimneys to deliver presents on Christmas Eve.

The *Star Trek* brand of teleportation technology – the transporter – was born out of a mundane problem facing the man who developed the television series in the sixties. Gene Roddenberry lacked the budget to show anything as dramatic as a starship landing each week. Non-fictional teleportation, however, has an impeccable pedigree originating in an intellectual punch-up between two scientific giants – Albert Einstein and Niels Bohr, the Danish father of atomic physics.

Einstein rejected the way quantum theory allowed ghostly 'action at a distance' in which one particle can influence another, regardless of distance. He also rejected the way that the laws of quantum theory allow atoms to make apparently random choices. The peculiar features of quantum theory do not end there. The theory holds that any measurement of a quantum property is liable to change that property. Indeed, quantum stuff does not even have properties independent of the act of observation. Atoms and electrons can be in two places at once, or behave in two apparently inconsistent ways simultaneously – both as particles and as waves. These split personalities obey equations that predict the results of subatomic experiments in terms of probabilities, not certainties. As one quantum pioneer ruefully remarked: 'You never understand quantum theory, you just get used to it.'

But Einstein refused to get used to it, and in 1935 fought against this assault on reality with a *Gedankenexperiment* – a 'thought-experiment' rather than one conducted in the laboratory. With two colleagues, Boris Podolsky and Nathan Rosen, Einstein noted that the theory applied not only to single atoms but to molecules made of many atoms as well. So, for example, a molecule containing two atoms could be described by a quantum expression – a wavefunction. Einstein realized that if you separated these two atoms, they would still be described by the same single wavefunction. In the jargon, they are 'entangled'. This has a strange consequence: the moment you measure something about one atom, the state of the other atom would instantaneously be transformed in a quite specific way – even if it was at the opposite end of the universe. That, said Einstein, would apparently violate his theory of relativity that states that nothing, not even information, can travel faster than light.

To conduct an experiment to investigate this paradox was beyond the technology of the 1930s. But three decades later, the Northern Irish physicist John Bell devised a test while working at CERN, the European particle physics laboratory in Geneva. And since the 1970s, John Clauser of the University of California at Berkeley, Alain Aspect at the Institut des Optics in Orsay, France, and others have been studying entangled particles. Their research shows that both common sense and Einstein were wrong. The action-at-a-distance that was derided as spooky by Einstein does indeed occur, and it is this peculiar feature of quantum theory that teleportation exploits.

The concept of action-at-a-distance relies on entanglement, enabling anyone who wants to carry out teleportation – including Santa and his scientific helpers – to side-step one key problem. To send a complete instruction set on how to remove Santa from his sleigh and rebuild him in the hearth of a chimney, the helper would need to measure the quantum state of every one of Santa's atoms. In other words, their energy, position, and state of rotation. But quantum theory dictates that any attempt to read the complete state would scramble it. This is a consequence of the so-called 'uncertainty principle' that lays down limits to what we can know about subatomic particles. The principle says that the more accurately you know the position of a particle, the less accurately you can know its speed at the same time. Similarly, if you know exactly where something is, you can know nothing about how fast it is travelling.

This idea, that won Werner Heisenberg the Nobel Prize for Physics, would seem to discredit quantum teleportation. Then, in 1993, IBM's Charles Bennett along with collaborators Gilles Brassard and Claude Crepeau at the University of Montreal, Canada, Richard Jozsa of the University of Oxford, Asher Peres of the Technion, Israel, and William Wootters of Williams College, Massachusetts, theorized that entanglement can overcome the uncertainty problem. Because in Einstein's experiment tweaking one atom means automatically tweaking the other, Bennett and his co-workers realized that the pairs of entangled atoms in effect made a 'quantum phoneline' across space. It should be feasible to 'teleport' the quantum state of one photon to another an arbitrary distance away. This opened up the possibility that a transporter could transmit atomic data. Zeilinger's team generated an entangled pair of photons by aiming a laser beam at a certain type of crystal. As a result, observation of one of the photons instantly altered the quantum state of the other – a feat that should be possible even if the two particles are separated by a cosmos,

The property the team teleported was polarization. In the

context of the wave picture of light, polarization refers to how light waves are restricted to certain directions of vibration. This is the same property that enables photons of one polarization to pass through a pair of sunglasses, unlike photons of the perpendicular polarization that are absorbed or reflected. During teleportation, an initial photon (that carried the polarization state to be transferred), and one of a pair of entangled photons were subjected to a measurement. This measurement entangled the initial photon with one in the pair, that itself was entangled – like a quantum daisy chain. As a consequence, a third photon – in the entangled pair located on the other side of the room – acquired polarization information from the initial photon. In other words, a quantum state was passed from one photon to another.

If the same arrangement were used to beam Kirk to a moon base, the *Enterprise*'s transporter would need a Kirk clone – not any old clone but one that is entangled with a second clone in a transporter on the surface of the moon. 'They are like identical twins who don't have a hair colour yet,' said Zeilinger. 'But as soon as you observe one, and he spontaneously assumes a hair colour, the other will adopt the same.' The first clone can be thought of as a blank sheet, actually a superposition of any quantum possibility, on which the teleporter 'writes' Kirk's quantum wherewithal. When Kirk asks to be beamed down, this clone is correlated with the second clone on the moon's surface, courtesy of entanglement. (Actually, it is even more complicated than this. To teleport Kirk, the *Enterprise* transporter must send additional information to the moon so that the second transporter can subtract the initial quantum state of the first clone, now mixed up with Kirk's vital quantum statistics, to leave the essence of 'Captain Kirkness' in the second clone.)

This experiment marked a key advance, although to find out if teleportation had occurred, Zeilinger had to destroy the teleported photon. It would be like mulching Captain Kirk, once he had successfully beamed down to the surface. This destructive problem was overcome in a more elaborate teleportation trial conducted in 1998 by Jeff Kimble of Caltech, along with Samuel Braunstein of the University of Wales at Bangor, and others. In effect, they used photons generated by a laser beam and a special kind of optical crystal to project the state of one photon on to another. Braunstein said that this marked the first true 'high-fidelity' demonstration of quantum teleportation (although Zeilinger points out that his teleportation experiments were equally high-fidelity – it is just that they only worked one time in every four attempts).

Remarkable as this research is, scientists have a long way to go before they can build anything like a *Star Trek* transporter. To beam Santa down to some point on the Earth, it is already clear you would require two entangled blank copies, or quantum clones. This is tricky. Leaving aside what Zeilinger describes as 'deep philosophical issues' to do with identity and distinguishability, you would also need to know Santa's complete quantum state, which is not just the energy state of every one of his component atoms but all their mutual entanglements, too. That is an incredibly huge task. Samuel Braunstein points out that you need about ten gigabytes (about 100,000 million bits, or ten to fifteen CD ROMS) to describe the full three-dimensional details of a human down to one-millimetre resolution in each direction. Scaling this to atomic resolution, you would need ten to the power of thirty-two bits (a one followed by thirty-two zeros). This is so much information that, even with the best optical fibres conceivable, it would take over 100,000 million centuries to transmit all that information. 'It would be easier to walk' he said. Indeed, when examining the physics of *Star Trek*, astrophysicist Lawrence Krauss calculated that it would take a longer time than the age of the universe to transport Mr Spock's vital statistics down to the surface of a planet.

However, Santa may have cracked some of the problems facing his onward delivery with the help of the next generation of computers that, intriguingly, also spins out of quantum theory and draws on some of the ideas investigated in teleportation research. The fathers of the modern computer, Alan Turing and John von Neumann, wrestled with the idea that computers must obey the laws of physics to provide a suitable physical foundation for the science of information-processing. Recently, Rolf Landauer, Richard Feynman, Paul Benioff, David Deutsch and Charles Bennett investigated the consequences of miniaturization and how micro-electronics are fast shrinking to the atomic realm ruled by quantum mechanics. After all, size does not matter to computers, because information is fungible – that is, it maintains its utility no matter what its physical form, or how small its basic units – a property not shared by shoes, for example. That means information can be carried by particles of light, electrons, or the spinning nuclei of atoms, all of which are ruled by quantum lore. This has crucial consequences that can be understood by comparing the abilities of a classical computer with those of a quantum machine.

The PC on your desk shuffles information in the form of binary numbers, those containing only the digits 1 and 0, that it remembers as the 'on' and 'off' positions of tiny switches, or 'bits'. By contrast, the switches in a quantum computer can be both 'on' and 'off' at the same time, in a superposition. These so-called 'qubits' could perform two calculations at once. Two qubits could thus do four things at once, three qubits could do eight, and so on. The more qubits, the more of a speed-up due to 'quantum parallelism' you get, leading to potentially mind-boggling performance. Rolf Landauer of IBM explains why this idea is so shocking: 'When we were little kids learning to count on our very sticky classical fingers, we didn't know about quantum mechanical superposition. We gained the wrong intuition. We thought that information was classical. We thought that we could hold up three fingers, then four. We didn't realize that there could be a superposition of both.'

The potential utility of quantum computers first became

clear when Peter Shor, a researcher at AT&T Laboratories, showed that they could crack codes much faster than conventional machines. Then Lov Grover of Bell Laboratories demonstrated their utility in sorting through lists, such as hunting for a name in a telephone directory, comparing diaries for a lunch date, or even finding efficient routes to deliver presents – the Travelling Santa Problem.

Progress has been made in developing quantum computers by storing information on charged particles held in a magnetic trap, on particles of light, and on 'artificial atoms' – tiny structures called quantum dots. At Stanford University, California, Isaac Chuang with Neil Gershenfeld of MIT took another important step by showing that quantum computing can also be carried out with ordinary liquids in a beaker at room temperature. This advance, also made by David Cory, Timothy Havel and Amr Fahmy at Harvard University, disproved the assumption that the quantum effects needed for computation could only be found in very cold, man-made systems.

The device was known as the 'coffee cup quantum computer'. Each molecule in the computer contains atoms, and the nuclei of atoms act like tiny bar magnets. However, they are not the ordinary bar magnets found on refrigerator doors. These magnets can only point in two directions, 'up' and 'down', because of a nuclear property called 'spin'. A single nucleus can therefore act as a qubit, its spin pointing perhaps up for 'off' and down for 'on'. A given spin lasts a relatively long time and can be manipulated with a technique called nuclear magnetic resonance, a feat performed by chemists for years to study the structure of molecules and now exploited by body scanners to study the structure of people. Thus each molecule can act as a little computer and is capable of as many simultaneous calculations as there are ways of arranging the spins in it, according to Chuang, now with IBM Research. He has even managed to perform some simple mathematical problems with chloroform molecules.

The implications of this work for Christmas science are all too obvious. Perhaps Santa uses a goblet of liquid for quantum computing. With the awesome computational power within a goblet of mulled wine or melted snowballs, Santa could handle endless e-mails and letters, hone the performance of his tele-porter, or crack the logistical problems he faces each Christmas Eve, when delivering presents to children across the planet.

EINSTEIN AND SANTA

Larry Silverberg at North Carolina State University takes a dif-ferent approach to answering all those troublesome questions about Santa's annual feat. Silverberg, a professor of mechanical and aerospace engineering and a member of NASA's Mars Mission Research Center, was assisted in his endeavour by Robert Stanley and Jeffry Stock-Windsor, doctoral students in mechanical engineering; J.P. Thrower, a doctoral student in electrical engineering; and Dan Deaton and Charles Grant, who are both seniors in mechanical engineering. Silverberg describes Santa's sleigh as one of the engineering wonders of the world and thinks he knows how it works, enough to become a true believer: 'Santa clearly is ahead of the curve when it comes to applying advanced scientific theories to his sleigh's design. Children shouldn't believe others who say he isn't real because there's no way he could deliver toys all over the world in one night. There is a way, and it's based on plausible science.'

Silverberg suggests that Santa exploits features of Albert Einstein's theory of relativity. The first part, which is con-cerned with so-called special relativity, starts from the premise that the speed of light and the laws of physics are the same – invariant – for observers moving at constant speeds relative to one another. The significance of special relativity is that common sense falls apart when we travel at high speed, particu-larly when we approach the ultimate speed, that of light.

However, the speed of light is constant, no matter what the position or speed of an observer.

That is why relativity seems so odd. A snowball thrown by Santa as his sleigh whizzes past would seem to him to move slowly compared with the earthbound perception of the snowball's speed. By contrast, from the vantage point of an observer on the ground, the snowball's velocity includes that of the sleigh. Relativity says that light breaks this commonsense rule. The speed of light from a torch turned on by Santa would therefore look the same to him, sitting in the moving sleigh, as to someone standing below in the snow.

In our everyday experience, it is time and space that are absolute: inches and seconds are the same, wherever you are on the planet or whatever you are doing. However, the only way observers in relative motion can come up with the same speed of light, say for a laser beam from a Christmas light display to pass from one planet to another in a given time, is if each of their 'seconds' or each of their 'inches' is different. The higher Santa's speed, the more time dilates and space contracts, says Larry Silverberg: 'You can see that it is quite easy.'

Special relativity thus gives Santa ample opportunity to deliver those presents in what for us is an eyeblink. To quote Silverberg again: 'In his reference frame, time moves much quicker than in our own – that is when he is inside his "relativity cloud." In the relativity cloud, because time moves much slower for us (relatively), he sees us basically frozen. He doesn't even need to hurry. He has all the time in the world.'

The next task for Silverberg and his team was to show how Santa achieves these huge speeds. Even if a rocket sleigh could get him to that small fraction of light speed, the fuel requirements would be prohibitive. This is where we need the second part of relativity theory, so called general relativity, which broadens Einstein's original theory to accommodate the idea that the laws of physics should be the same for all observers, regardless of how they are moving relative to one another.

General relativity came about after Einstein realized that if Santa Claus fell off a roof he would not feel his own weight – until he hit the ground, that is. The theory replaced Newton's way of describing gravity, seeing it not as a force but as the curvature of 'spacetime', a four-dimensional mix of space and time. For example, the Earth orbits around the Sun because spacetime has been warped into a shape like the inside of a bell, with the Sun sitting at its centre.

Silverberg points out that although our theoretical understanding of spacetime was laid down during the first decades of this century, 'Santa and his community at the North Pole have known about this much longer. In fact, they have learned, as a practical matter, how to manipulate time, space, and light – to the extent that they can control these phenomena.' We have only begun to appreciate the ramifications of relativity. 'But Santa's society, with its hundreds of years of experience manipulating and understanding relativity, has turned it into a working theory – they've made relativity clouds that fit Santa, his sleigh and all of the reindeer.'

According to relativity theory, matter cannot move through space faster than the speed of light. However, there is no limit on the speed with which space itself can move: the sleigh can sit at rest in a small bubble of space that flows at 'superluminal' velocities through normal space. Another way to put it is that objects can travel locally at slow velocities yet be travelling faster than the speed of light because the very fabric of spacetime is stretching.

In 1994 Miguel Alcubierre at the University of Wales College of Cardiff demonstrated that this apparent loophole is entirely consistent with modern physics. Alcubierre found that if spacetime can locally be warped so that it expands behind Santa's sleigh and contracts in front of it, then the craft will be propelled along with the space it is in, in effect riding the crest of the wave. By this scheme, the spacetime around the sleigh is being warped so that it can be moved between chimneys without experiencing much, if any, local acceleration; instead, the spacetime between Santa and the last chimney is expanding, bringing the next closer. The sleigh will never travel locally faster than the speed of light, because the light, too, will be carried along with the expanding wave of space. However, the accompanying 'wave' will allow huge distances to be travelled in almost no time at all, let alone the short distance between chimneys.

Santa can go even further and cut and paste different bits of the universe together, using spacetime short cuts called 'wormholes', an idea explored by the mathematician Ian Stewart at Warwick University in England. If you imagine spacetime as being curved, like a sheet, a wormhole offers a more direct route than following all the curves and contours of the sheet – one that goes outside the normal universe. Santa would enter one portal, pass through the wormhole and emerge from another. Mr Claus can carry one end of the wormhole on his vehicle, and arrange for the other to materialize inside each dwelling that he visits. No more sooty chimneys or getting stuck inside central heating systems, says Stewart.

The wormhole idea can also allow time travel, providing unlimited opportunities to deliver those presents. This possibility was raised by work published in 1988 by Michael Morris, Kip Thorne and Ulvi Yurtsever and rests on the so-called twin paradox. Imagine two identical reindeer, Donner and Blitze, says Ian Stewart. Donner remains on Earth and Blitze heads off into space at nearly light speed, returning forty years later as measured by an earth-bound observer. Donner has aged forty years, but because of time dilation Blitze has aged only a few. Morris, Thorne and Yurtsever realized that by combining a wormhole with the twin paradox, they could get a time machine. The idea, explains Stewart, is to leave one end of the wormhole fixed, and to zigzag the other end to and fro at just below the speed of light. Seen from inside the wormhole, both ends age at the same rate. But from outside, the moving end ages more slowly because of its speed.

From the perspective of an elf hanging about the fixed wormhole portal, therefore, a day-long journey of the moving end could appear to take ten days. But if he were to peer through the wormhole, the curious elf would see things as they were nine days earlier. This means that the passage of time is different if you wriggle from one end of the wormhole to the other. Indeed, if you travel through normal space to the moving end and then dive through the wormhole, you end up in your past.

There are still some problems to overcome. For example, time travel risks creating paradoxes that defy commonsense ideas about cause and effect. The classic one is the grandfather paradox, where Santa goes back in time and runs over his three-year-old grandfather with fatal consequences for Santa himself. One way to side-step such paradoxes may be offered by the many-worlds interpretation of quantum mechanics, the theory that rules the subatomic world. This controversial interpretation of quantum mechanics is, in the words of one critic, cheap on assumptions and expensive on universes.

If you believe it (and many don't), each time shift carries us into a new version of the universe, co-existing with the original but separated from it along some totally new kind of dimension. There could be trillions of Santas out there delivering presents in parallel universes. However, this presents us with a new problem. In all Miguel Alcubierre's talk of weird and warped spacetime allowing Santa to travel at light speed and beyond, we have overlooked how Santa manages to mangle the fabric of the universe in the first place.

To warp spacetime would require the presence of exotic matter that, unlike the stuff we are familiar with, repels other matter by exerting an anti-gravity effect, according to Alcubierre. That is, as long as we adhere to the theory of relativity. Quantum mechanics does allow 'negative-energy' matter on the scale of atoms and molecules. The big question is whether matter can have this property on the scale of reindeers and sleighs. A recent calculation investigated this very issue. Mitchell Pfenning and Larry Ford at Tufts University in Massachusetts found that this energy would have to be packed into a doughnut-shaped region wrapped around the sleigh. However, they calculated that the total amount of energy needed to sustain the warp is huge – around 10 billion times the energy locked up in all the visible mass in the Universe. That would seem to rule out warp-drive sleighs.

But it may not be over for Santa, according to Alcubierre. Pfenning and Ford's work is based on approximations which, strictly speaking, only work in space that is not already warped. At present, we lack the theory to do this kind of calculation, a fusion of quantum mechanics and relativity. Has Santa learnt how to exploit this next generation of theory, so-called quantum gravity? Given the evidence of his activities each Christmas Eve, this may be the explanation.

THE TRAVELLING SANTA PROBLEM

There are many other questions left to tackle, however. How does Santa know where children live, and what gifts they want? Although old-fashioned letters to Santa still work, Silverberg and his students speculate that Santa relies on a strategically placed multigrid-receiving antenna system that picks up electromagnetic signals from children's brains. There is, after all, a technique called magnetoencephalography which uses a SQUID (a Superconducting Quantum Interference Device) to detect minute magnetic fields generated by the crackle of brain activity. Sophisticated signal-processing methods could then be employed by Santa to filter the data and ascertain who the children are, where they live and whether they've been bad or good. This data is transferred to Santa's onboard sleigh-guidance system, which uses a computer software program to plan the most efficient route of delivery.

Here, of course, he runs into the classic dilemma faced by

the travelling salesman who has to visit a number of cities, say N, each only once and in such a way as to minimize the total distance travelled: he has a penny-pinching boss and has to keep fuel costs as low as possible. In the festive season, many scientific minds turn to solving that perennial chestnut, the Travelling Santa Problem. For a handful of cities and roads, it may be easy to determine the shortest route because not that many options exist. If the number of cities is five, say, a computer could easily calculate the 120 possibilities. With ten cities, there would be 3,628,800 possibilities.

However, even with the number-crunching power of the fastest available machine, the time required to solve the problem rapidly spirals out of control. For just twenty-five cities the number of possibilities is so immense that a computer evaluating a million possibilities per second would take 490 billion years, that is about forty times the age of the universe, to search through them all. For the 842 million households, he would take 10 to the power of 7.15 billion years. Now that presents Santa with a headache, if he is trying to be efficient.

Larry Silverberg points out that present-day computers could calculate a route that is 99.99 per cent 'near' the best, which would be sufficient for Santa's purposes. The Travelling Santa Problem is broken down. Cities are grouped into countries and weighted according to the number of children in the country. Then each city is given a weighting number which corresponds to the number of children in that city, and so on. 'We start by optimizing by country and then by city and finally by home. We can perform more subdivisions when needed, but in the end the number of calculations can be performed using present-day, albeit fast, computers,' Silverberg explains. 'When you call a friend, telephone companies use switching algorithms that are related to these kinds of "suboptimal solutions."'

Given Santa's extraordinary research-and-development operation, he probably has a quantum computer. He may even possess a DNA computer, an idea pioneered by Leonard Adleman of the

University of Southern California. In a milestone paper, Adleman tackled something called the 'directed Hamiltonian path problem', which also involves finding a special path through a network of points. We can think of this in the same terms as the problem facing Santa. The computer code consisted of a mixture of trillions of pieces of single-stranded DNA, each piece either representing a city or a route. Among the vast number of combinations that resulted from the binding together of complementary DNA strands, it was overwhelmingly probable that one combination corresponded to the solution sought. Standard molecular-biological techniques were then used to fish that molecule out: the 'solution' could be distinguished by its length and details of its composition. While current supercomputers are able to perform a million million million operations per second, molecular computers could conceivably run billions of times faster. That is fast enough, when combined with mathematical cunning, for Santa to plan the most efficient route.

But Santa's challenges are not all over. How can one sleigh carry 2 million tons of presents? It doesn't have to, says Larry Silverberg. To tackle this problem, one dear to the heart of every child, he invokes nanotechnology, an idea first put forward by the late Nobel laureate Richard Feynman. Santa can now use a hierarchical distributed mobile manufacturing system to make the gifts on site in each child's home. These silicon-chip-based machines, so tiny they can fit on the head of a pin, are loaded with a code containing the child's toy list. The machines literally grow the toys, atom by atom, from bits of snow and soot Santa collects along his route. Large toys require thousands of nanomachines working in concert and can drain Santa's technological resources, Silverberg warns, which is why children shouldn't expect more than one big gift each year.

DNA computers, nanotechnology and warped spacetime seem fair enough, but how does a reindeer fly? It's in their genes, says Larry Silverberg. After centuries of selective breeding and (more recently) bioengineering, their lungs are such that, when filled with an appropriate mixture of helium, oxygen and nitrogen, they become buoyant. Flying while pulling the sleigh is then as easy for them as pulling a raft in water. For extra speed, the reindeer could always be powered by jetpacks, adds Silverberg. He cites a soccer game that was 'kicked off' with a mascot strapped to jetpacks who touched down in the middle of the stadium. 'In fact, the reindeer don't even have to be lightweight – although it helps. Our team at NC State speculates that Santa uses reindeer simply because they are his favourite Arctic animal. Also, they have good balance, which is needed when landing on rooftops.'

Despite this natural ability, the reindeer may need to develop stabilizers to keep Santa, the sleigh and all those presents on an even keel, and this is a genetic possibility. In studies of fruit flies, scientists have discovered so-called homeotic genes carrying a mutation that, amazingly, can transform one body part to another, such as an antenna to a leg, a thorax to a wing, a nose to a leg, and so on. Reindeer could be customized for flight by manipulating these homeotic genes, speculates Matthew Freeman, a geneticist at the Laboratory of Molecular Biology in Cambridge: 'Work by Cliff Tabin of Harvard Medical School and others has shown that genes that specify limbs in flies, such as wings and legs, are the same as in mammals and even work in much the same way. That means that, with Santa's genetic wizardry, Rudolph could grow stabilizers or even wings.'

Ian Stewart adds that we may have underestimated the importance of one of Rudolph's most prominent features – his headgear. He points out that the aerodynamics of hypersonic flight is highly counter-intuitive, anyway, though perhaps no more so than the good old heavier-than-air flying machines we now take for granted. 'Reindeer have a curious arrangement of gadgetry on top of their heads which we call "antlers", and naively assume exist for the males to do battle to win females,' he says. 'This is absolute nonsense. The "antlers" are actually fractal vortex-shedding devices. We are talking not aerodynamics here, but antlaerodynamics.'

At supersonic speeds, Concorde generates lift by shedding a big helical vortex from the tip of each wing. The reindeer, at their much higher speeds, play the same trick with the prongs of their antlers, explains Stewart. 'These spin off a whole system of vortices, carefully tailored to generate the right amount of lift for such high speeds. (You don't need much!) So the reindeer hang from their antlers as they fly – which is why the antlers are on top, and at the front.'

No doubt Santa used supercomputers to design the optimal antlaerodynamic configuration, and genetic engineering to produce it in the actual reindeer. Just as penguins look comical on land but come into their own underwater, so the apparently earthbound reindeer only reveals the true beauty of its design when travelling at speeds in excess of Mach 6000. But, of course, if the reindeer travel too fast, they could burn up, along with the sleigh. To provide adequate thermal shielding, both of reindeer and warp-drive sleigh, Larry Silverberg's team at North Carolina proposed a Kevlar-like composite fibre encapsulated in an epoxy resin matrix, making it very strong, lightweight, durable and cold-resistant. Other possibilities include the silica materials used on the Space Shuttle.

Anecdotal evidence also backs the idea that Santa uses heat shields, says Silverberg: 'When going through the ionosphere, a sleigh made of this material would glow like a speck of bright red, a sight children and other reliable sources have long reported seeing in the sky on Christmas Eve.'

The US Air Force 48th Fighter Wing claims to use satellite dishes to track Santa on Christmas Eve, with other Air Force Space Command squadrons around the world, to prevent the unnecessary scrambling of interceptor aircraft and ensure the safe arrival of 'the Jolly Old Elf' and all his presents. 'We have some of the most sophisticated equipment in the world. The deep space tracking system was constructed at a cost of over

$600 million. Santa is in good hands,' said Tech. Sgt. Ray Duron, Crew Chief of the 5th Space Surveillance Squadron at RAF Feltwell, which coordinates the route of his sleigh with the 1st Command and Control Squadron in Colorado Springs.

Given the extraordinary array of technology already used by Santa, much of which is beyond the capabilities of the US military, this annual 'Santa Track' – which dates back to 1957 – seems unnecessary. Indeed, some might say it is merely a publicity stunt engineered by defence scientists to draw attention away from the vast range of scientific and technological achievements pioneered by Santa to ensure children across the world are not disappointed on Christmas morning.

twelve

Christmas 2020

'Ghost of the Future!' he exclaimed, 'I fear you more than any Spectre I have seen. But as I know your purpose is to do me good, and as I hope to live to be another man from what I was, I am prepared to bear your company, and do it with a thankful heart. Will you not speak to me?'

Charles Dickens, *A Christmas Carol*

I was awoken by a shaft of light from my beside lamp. Without a dose of daylight-intensity photons, I would have been even more depressed about Christmas Day than usual. The flatscreen TV hanging on the wall crackled into action to give me weather diagnostics, digital yuletide greetings courtesy of a cola manufacturer, and a glimpse of the world outside my 195th-floor apartment. The snow lay deep and crisp and even, just like the atmospheric engineers had promised. 'Silent Night' was blaring from a passing Santa-shaped dirigible, whose digital hoardings exhorted the masses to finish their last-minute 'teleshopping'. And an e-mail told me the local transgenic farm had delivered a 3-lb hi-fibre cloned turkey breast fillet, engineered with a mixture of tenderizing enzymes and Maillard reagents to improve its taste and colour.

A hum sounded nearby. My 3D-fax was delivering another gift sent over the web. Once I had visited my sanichamber and gargled with my cinnamon-flavoured, plaque-digesting enzymes, I decided to take a peep. Through the quartz-glass window of the 3D-fax I could see sparkles of red, blue and green light as laser beams crossed in the vapour, building up polymer deposits into another present. I yawned. It looked as if it was going to be one of those designer-Santa effigies you could personalize with the sender's or recipient's face.

Fortunately, the neural net software in my message centre had already figured out whom the gift was from, assessed its value to the last euro, and used a game theoretic software package to select the appropriate reciprocal gift to send back via the 3D-fax. All that, and a personal message of thanks, lovingly crafted by fuzzy-logic software, then signed with the appropriate seasonal wishes – Happy Kwanzaa/Christmas/Hanukkah/Holidays. Thank goodness we don't have to think about what to give people any more.

I put on my virtual-reality headset to check out the latest store offers before lunch, guided by a small herd of shopbots that had spent the past few hours grazing on the internet. Click on, cruise through a mall web site and pile your cyber-shopping trolleys high with the latest techno-magic! Experts used to predict that Christmas shopping at virtual stores would be a lonely affair. But now that everyone could cruise around as an avatar, a digital alter ego, it was fun to see the last-minute frenzy without all the physical hassle, trolley rage and fights over the last remaining turkey or mince pie.

Look at that. A final discount on cloned firs was broadcast. They were the latest Gucci trees, tear-drop shaped with a designer logo *and* a guarantee that not a single GM needle would fall upon your self-cleaning floor. Two eyeblinks and a nod of the head later, I started to examine the goods. I breathed in. The odourchip in the headset sensed this sharp intake of conditioned air and did a passable imitation of the tree's smell, down to the last whiff of alpha pinene. 'I'll take it,' I said.

After the usual KaDeWe quantum-encrypted credit checks and offer of GM extras (no, I did not want the tree's jellyfish gene activated – glowing trees are so vulgar), my customized Gucci tree dropped into my trolley, complete with black fairy. Twenty thousand euros, a three-hour delivery from tissue plant to door, plus a trade-in for my old biodegradable plastic tree. It was a bargain.

Off with the VR headset and into the kitchen. The turkey had already been screened for harmful bacteria with a GeneChip and would only take a few minutes to deal with. The smart laser cooker had selected the appropriate vegetables, seasoning and so on. The 'Fat Wallaby' Shiraz was on the table, ready to drink. This time I had made sure the wine would not interfere with any mood-altering drugs by selecting one of the low-alcohol engineered reds from Australia. Armed with my body mass and fat index, updated from my brief visit to the sanichamber, the small pack of tabletop electronics in the Dietmate had already worked out what would be healthy for me to eat, and started to mutter something about cutting the calories by boiling the potatoes. I decided to overrule it today and issued the command: 'Roast them in dripping!' The Dietmate began to plan low-fat retribution for Boxing Day.

Later came the highlight of the celebrations – a virtual Highfield family reunion. This year we had consulted an imagineer at Disney's virtual worlds, one of the best-known providers of on-line gatherings. For our get-together we had asked for 'something like Dingley Dell *and* Santa's grotto' with the full effects – smellyvision, virtual celebs, the works. There were the usual jokey avatars. I went for Thomas Nast-style Santa garb and a seasonal cinnamon smell, and selected a winged Rudolph to get me around. The motion-feedback chair made me feel a little seasick when my reindeer started to fly.

There were other surprises: photos of some long-dead, much-loved relatives, scraps from their voice mail and a few family anecdotes cobbled together by the imagineer and processed by computer. Though long gone, their digitally remastered bodies lived on to send a brief Christmas message and even answer a question or two at the family reunion. Using black and white images and a crackly soundtrack, the effect was eerie. I bet a few VR headsets misted up after that.

Earlier that week the family elders had spent some time telling the imagineer who got along with whom, details of the worst family feuds, that sort of thing. Then the high-level software could distract us with a virtual celebrity, even march us off with a virtual elf if there was any risk of clash between avatars. On-line feuds can be much worse than the real thing. Once you have rowed in cyberspace, you can't easily forgive and forget. Victims replay and brood, replay and brood until the next Christmas. Just in case there were any clashes, we took the usual pharmacological fixes – pheromones, dopamine boosters, serotonin blockers – to tweak brain receptors and neurotransmitters for the full festive mood.

And, of course, we selected the digitally customized film version of *A Christmas Carol* for those who wanted to attend but sought refuge from some of the relatives. There are a lot of options available in these films – too many. This had prompted the usual flurry of e-mail, voice mail and pic mails in the preceding days. We all agreed on a period setting, then squabbled about who should play whom. In the end we compromised with Alec Guinness as Scrooge and Stan Laurel playing a particularly pathetic Bob Cratchit. The supercomputer did the rest. For the sound track we had data-mined one of the entertainment archives, using a neural net to cross-reference our personal tastes with appropriate music. Then all we had to do was settle down to watch in our VR headsets. A synthetic orchestra and choir began to play a carol and a ghostly image of Marley appeared. As I sat under my cloned tree, I swallowed a smart pill to try to help unlock some of my childhood memories. The snow, seeded with silver iodide, continued to fall outside. I felt a warm glow inside. You can't beat a traditional Christmas.

appendix one
The Formula for Christmas Day

For all those who like to plan their Christmas in advance, here is a little formula that can reveal the day of the week on which Christmas Day falls in any year (including leap years).

1 Write down the year you're interested in, say 1998. Split this number into its century number C (= 19, in this case) and its year number Y (= 98).

2 Now, divide C by 4, and keep just the whole-number part of the result, which we will call K. Thus in our case, K = 19 divided by 4 = 4.75, which is rounded off to give K = 4.

3 Do the same for Y, again keeping just the whole number, giving a new figure, G. In this example, G = 98 divided by 4 = 24.5, rounded off to 24.

4 Now work out the value of D, using the formula D = 50 + Y + K + G − (2 × C). In our case, this gives D = 50 + 98 + 4 + 24 − 38 = 138

5 Finally, to work out the day of the week on which Christmas falls, divide D by 7, and write down the remainder, R. Use the following table to give the day of the week:
    R = 0 Sunday
    R = 1 Monday (and so on)
    R = 6 Saturday
In our case, D = 138, so D divided by 7 is 19, with a remainder (R) of 5. This means Christmas Day 1998 fell on a Friday. The

same formula shows, for example, that Christmas Day during the year 2164 will fall on a Tuesday. Note that the formula applies to the Gregorian calendar, first introduced in 1582 and still in force.

Is Faith Good
for You?

There is growing evidence that those who respect religious traditions, and presumably those who take Christmas seriously, can expect a healthier life.

A wide-ranging survey of scientific evidence of the 'faith factor' in disease has been conducted by Dale Matthews of the Georgetown University School of Medicine. He reviewed 212 studies and found that religion had a positive effect in three-quarters of them, notably those investigating substance abuse, alcoholism, mental illness, quality of life, illness, and survival. 'Whether one looks at cancer, hypertension, heart disease, or physical functioning, most of the studies demonstrate a positive effect,' he said. Seven out of ten studies on survival showed that religious people live longer.

Among the investigations Matthews cited are the following:

● A survey of 91,909 individuals in Washington County, Maryland, found 50 per cent fewer deaths from coronary artery disease, 56 per cent fewer deaths from emphysema, 74 per cent fewer deaths from cirrhosis, and 53 per cent fewer suicides in religious people.

● A landmark study of 4,725 people in Alameda County, California, showed that church members had lower mortality rates than others, independent of socioeconomic status and smoking, drinking, physical inactivity, and obesity.

● A study of 522 Seventh-Day Adventist deaths in the Netherlands revealed they had an additional life expectancy of nine years for men and four years for women when compared with the general population. Although the group is vegetarian, Matthews added that 'the reason that they are vegetarian is their religion.'

● Mormons enjoy unusually good health, with cancer and heart disease rates less than one-half those of the general population. Highly religious Mormons in Utah experience one-half the rate of cancer, as do less adherent members of the faith, even though they constitute a highly homogeneous social group.

● Among 1,344 outpatients in Glasgow, those who participated in a religious activity at least monthly were less likely to report physical, mental, and social symptoms.

● A study of 1,400 women in Michigan found that church attendance was correlated with a longer life, while 'TV watching was associated with death', said Matthews. 'This presumably is a variable for lack of physical activity, but you never can tell.'

● And almost all patients who undergo heart surgery – including atheists – pray. A study of bypass surgery patients showed that the six-month mortality rate was 9 per cent overall, 5 per cent in churchgoers, 12 per cent in those who did not go to church, and zero in the 'deeply religious.'

Since Matthews's roundup, there has been other support for the faith factor, notably two reports, coauthored by Ellen Idler, at Rutgers's Institute for Health, Health Care Policy, and Aging Research, and Stanislav Kasl, of Yale University School of Medicine. The papers, published in the *Journal of Gerontology*, reveal the findings of a twelve-year study that sampled 2,812 people age sixty-five and over from Protestant, Catholic, Jewish, and other religious backgrounds in New Haven, Connecticut.

The first study found a lower frequency of unhealthy behaviors; better support systems and social ties; and improved emotional well-being. The second report explored how religious involvement can influence changes in physical health over a twelve-year period, revealing that attendance at religious services was a good predictor of functional ability in later life.

'Over the long term, people who had better health levels in 1982 and continued attending religious services were able to maintain higher levels of functioning, and psychological health through 1988,' Idler remarks. 'Even after we took out the other variables such as friendship, leisure activities, and social support, there was still evidence that attendance at religious services had a positive impact.

'There were so many reasons for thinking that we should expect better health among people who are religiously involved, but until now it wasn't anything we were able to quantify,' she says. 'We also found that it wasn't a person's individual feelings of religiousness that made the difference; it was acting as part of the larger worship group that fostered positive health.'

Patients are to some extent driving this interest in the link between faith and health. A 1994 study found that 73 per cent of hospital inpatients prayed on a daily basis, 77 per cent said that doctors should consider their spiritual needs, and 48 per cent wanted their doctors to pray with them.

Doctors, however, are cool to the idea of the faith factor, according to Kenneth Pargament of Bowling Green State University. A study of 2,348 psychiatric surveys conducted over five years found that few collected basic data on religious belief: only 2.5 per cent used a religious measure and only 0.1 per cent had religion as a central variable, with only one employing a validated measure of religious commitment. Pargament goes so far as to argue that health research that ignores religion is incomplete. 'Faced with the insurmountable, the language of the sacred – hope, surrender, forgiveness, serenity, divine purpose – becomes more relevant,' he says. 'Ultimate control is still possible through the sacred when life seems out of control.'

One reason that doctors are not swayed by a simple picture of the 'faith factor' is that certain forms of religious expression are clearly counterproductive for better health. One has only to think of the tragedies surrounding the Jonestown, Waco, and solar temple cults. In addition, some groups reject medical care, such as the Faith Assembly in Indiana, which has three times the state average perinatal mortality rate and one hundred times the maternal mortality rate because of lack of obstetric care.

'When you look more closely, you find there are certain types of religious expression that seem to be helpful and certain types that seem to be harmful,' Pargament says. In several studies involving hundreds of subjects, he has found that people who embrace what could be called 'the sinners in the hands of an angry God' model do indeed have poorer mental health outcomes. People who feel hostility toward God, believe they're being punished for their sins, or perceive a lack of emotional support from their church or synagogue typically suffer more distress, anxiety, and depression.

By contrast, people who embrace the 'loving God' model see God as a partner who works with them to resolve problems. They view difficult situations as opportunities for spiritual growth. And they believe that their religious leaders and fellow congregation members give them the support they need. The result? They enjoy more positive mental health outcomes, Pargament says.

Support for this view has come from the work of Lee Kirkpatrick, at the College of William and Mary, who discovered that people who viewed God as a warm, caring, and dependable friend were much more likely to have positive outcomes than people who viewed God as a cold, vengeful, and unresponsive deity or who weren't sure whether or not to trust God. 'People who classified their attachment to God as secure scored much lower on loneliness, depression, and anxiety and much higher on general life satisfaction,' Kirkpatrick explains.

There is a deeper issue here. What, then, determines what we believe? Lewis Wolpert of University College London argues that the concept of cause was fundamental to human evolution and was a key driving force, for it allowed our ancient ancestors to construct complex tools. In turn, tools and other technology gave them their particular advantage over many other animals.

Yet there were many important events for which our ancestors could find no clear cause. Infants died of disease. Volcanoes, lightning, and natural disasters abounded. 'But they had no explanation for such distressing occurrences. They must have felt uncomfortable about their inability to control or understand such events, as indeed many still do today. As a consequence they began to construct, as it were, false knowledge. Indeed I argue that the primary aim of human judgement is not accuracy but the avoidance of paralysing uncertainty.'

This gave our ancestors two advantages that enabled them to adapt to a tough environment: uncertainty, and thus anxiety, was removed, and there was a God, spirit or other agent that might be appeased by a dance, offering or sacrifice. 'Might it not be that those with this disposition of thought survived better than those who did not have such beliefs?' he said. 'If that was the case, any genes linked with a propensity to believe would come to dominate in future generations.' Is this why so many people believe in Santa Claus?

Glossary

ABSOLUTE ZERO   No Christmas can get colder than this. The lowest temperature that anything can approach is $-273.15°C$ or $-459.67°F$ (for comparison, a deep freezer is around $-15°C/5°F$). A fundamental law, known as the third law of thermodynamics, says that nothing can be cooled down to absolute zero, although scientists can now come to within a few billionths of a degree of this baseline. As something cools, the average energy and movement of its molecules drop. You might think they would come close to stopping altogether near absolute zero, but quantum theory says that even at such extremely low temperatures, molecules cannot help but jiggle about a bit.

AMINO ACIDS   The molecular building blocks of proteins.

ATOMS   Regarded by the Greeks as indivisible units of matter, atoms are now known to be the smallest units to bear the chemical characteristics of an element, whether hydrogen or uranium. Each atom consists of a positively charged nucleus orbited by a mist of negative charge. They are mostly empty space: the nucleus, where most mass resides, is 100,000 times smaller than the overall atom. More than a billion atoms would fit on the full stop at the end of this sentence.

CELL   A discrete, membrane-bound portion of living matter, the smallest unit capable of an independent existence.

CHAOS   A term often used to describe preparations for Christmas. In science, it denotes apparently random behaviour and is the subject of a branch of mathematics called chaos theory. The essence of chaos is expressed in the butterfly effect: a butterfly flapping its wings near Santa's grotto can cause a subsequent hurricane over Texas. Chaos theory has put paid to the prospect of truly long-range weather forecasting, since the Earth's atmosphere is so 'sensitive' that if there is even the slightest uncertainty in the current weather conditions, then predicting the weather in a few weeks' time, such as a white Christmas, becomes impossible.

CHRISTMAS 'A day set apart and consecrated to gluttony, drunkenness, maudlin sentiment, gift-taking, public dullness, and domestic behavior,' according to the American writer Ambrose Bierce.

CHROMOSOME A long strand of DNA, containing a package of thousands of genes. There are twenty-three pairs in all human cells, except eggs and sperm.

COMPUTER A device that operates on data (input) according to specified instructions (program), usually contained within the computer and producing results. In a digital computer all values are represented by discrete signals – such as ones and zeros – rather than by continuously variable values, as found in an analogue machine.

CROSSLINKING The process that makes a turkey tough, involving the formation of side bonds between different chains in a polymer, such as the proteins in meat.

DETERMINISM The theory that a given set of circumstances inevitably produces the same consequences, rather like exuberant office parties and hangovers.

DNA (deoxyribonucleic acid) The vehicle of inheritance, from reindeer to wise men. A chain-like molecule made up of a series of 'bases' that come in four flavours (adenine, guanine, cytosine and thymine, or A, G, C and T), DNA carries the genetic blueprint for the design and assembly of proteins, the basic building blocks of life. This blueprint is provided by the order of the bases, which form a three-letter code (for example, ATT) for a particular amino acid that, when joined with a string of others, makes a protein. Santa, like all humans, has about 3,000 million bases in his genetic makeup, or genome, but only around 30,000 genes that work to make proteins.

ENTROPY   A quantity that determines a system's capacity to evolve irreversibly in time – for instance, the propensity of snowflakes to melt. Loosely speaking, we may also think of entropy as measuring the degree of randomness or disorder in a system.

ENZYME   Huge proteins that act as biological catalysts, accelerating essential chemical reactions in living cells. The name comes from the Greek words for 'in yeast'.

EVOLUTION   From the Latin *evolutio* ('unfolding'). The idea of shared descent of all creatures, humans, reindeer and robins included. Proposed in its modern form by Charles Darwin (1809–82) and described by him in *The Origin of Species* (1859). It is based on inherited diversity (variants that either increase their carrier's ability to make copies of themselves, and hence become more common, or hinder it, and become more rare) and natural selection (a struggle for existence that means that not all those born can survive and pass on their heritage). In time, Darwin suggested, this process has led to the evolution of new forms of life – the origin of species. Darwin's ideas fit perfectly with those of modern genetics, which holds that diversity arises through mutation, or random changes in DNA, from generation to generation.

FRACTAL GEOMETRY   From the Latin *fractus* ('broken'). The geometry used to describe an irregular shape that appears the same on all scales, whether you look at it close up or far away. Fractals display the characteristic of self-similarity, an unending series of motifs within motifs, repeated at all length scales. They abound in nature; examples include clouds and snowflakes.

FULLERENES   Discovered in 1985, this new form of carbon, distinct from carbon and graphite, amazed chemists because carbon compounds had been intensively studied for decades without a hint of its existence. The most famous fullerene, found in flickering candle flames, is buckminsterfullerene, or the buckyball, a molecule consisting of sixty carbon atoms. These football-shaped molecules, with twenty hexagonal surfaces and twelve pentagonal surfaces, reminded researchers of the geodesic domes designed by American architect R. Buckminster Fuller, hence their name.

FUZZY LOGIC   In mathematics and computing, a form of knowledge representation suitable for imprecise notions such as 'cold', 'loud' and 'Christmassy' that depend on their context.

GAME THEORY   A branch of mathematics that deals with strategic problems, such as those that arise in business, commerce, evolution and warfare, by assuming that the organisms involved invariably try to win. It can equally well be applied to Christmas shopping.

GENE   A unit of heredity comprised of the chemical DNA, responsible for passing on specific characteristics from parents to offspring, such as a propensity to pile on the pounds when overeating.

GENETIC CODE   The sequence of DNA bases, which spells out the instructions to make AMINO ACIDS.

GENETIC ENGINEERING   Tinkering with the GENETIC CODE to produce animals and plants with desirable properties – for example, Christmas trees that don't shed their needles so easily.

GLUCOSE   A sugar present in blood that is the source of energy for the body.

HO-HO-HO   Expression of mirth that begins as a deep breath. There follows a series of spasmodic involuntary outward expirations, controlled by the opening and shutting of the glottis, the pathway from the throat to the lungs. The muscles then act on the larynx, lengthening or shortening the vocal cords to create different tones out of the explosions of air. The head and neck together act as an instrument to modify the sound.

IMMUNE SYSTEM   The range of weapons at the body's disposal to fight invaders such as bacteria, viruses and fungi.

INTRACTABLE   Used to describe a problem that is so demanding that the time required to solve it rapidly spirals out of control. One example is the Travelling Santa Problem, in which our fat friend has to work out the most efficient route by which to deliver all those presents on Christmas Eve.

IONOSPHERE   The layer of the Earth's atmosphere, between 61 and 1000km (38 and 620 miles), that contains sufficient free electrons to modify the way that radio waves are propagated.

MAGNETIC RESONANCE IMAGING (MRI)   A noninvasive and painless scanning technique for examining and depicting the interior of the body, including its metabolism. It can also be used to find a sixpence in a Christmas pudding.

MAILLARD REACTION   A chemical reaction between carbohydrates and the amino acids of proteins that is responsible for the browning of certain cooked foods, such as turkeys.

MEGA   Prefix denoting multiplication by a million.

MOLECULAR BIOLOGY   The study of the molecular basis of life, including the biochemistry of molecules such as DNA.

MUTATION   A change in genes produced by a chance or deliberate change in the DNA that makes up the hereditary material of an organism. For example, a mutation in a gene may be one reason why Santa is so fat.

NANOTECHNOLOGY   The building of devices on a molecular scale (nanos is Greek for dwarf). Some believe that Santa exploits this technology to make presents.

NEURAL NETWORKS   Computers that can learn and are loosely modelled on the vast interconnected networks of nerve cells (neurons) in the brain.

NEURON   The nerve cell that is the fundamental signalling unit of the nervous system.

NEUROTRANSMITTER   A chemical that transmits signals between nerve cells. A drop in the rate of neurotransmitter release may be the reason for slurred speech, clumsiness, slow reflexes and loss of inhibitions in seasonal revellers who have consumed too much alcohol.

NUCLEIC ACID   A complex organic acid that forms the basis of heredity. Made up of a long chain of units called nucleotides and found in two types, known as DNA and RNA.

OCCULTATION   The passage of one astronomical object in front of another, obscuring it, as, for example, when the Moon moves between the Earth and Jupiter, hiding it from view. This kind of event may have been the portent the Wise Men were looking for to signify the birth of Jesus.

PARTHENOGENESIS   Reproduction without sexual union.

PHASE CHANGE   A change in the physical state of a material from solid snowman to liquid water, water to vapour, or solid to vapour.

PHEROMONE   A chemical substance secreted by an animal which influences the behaviour of other animals.

PHOTOSYNTHESIS   Probably the most important chemical reaction on earth, enabling Christmas trees and other plants to trap energy from the sun and convert it into a form that can sustain living cells. The key product of photosynthesis, which takes place in structures called chloroplasts, is the chemical adenosine triphosphate, the fuel for all the cell's activities. Eventually, that energy fuels creatures higher up the food chain, such as human beings.

POLYMER   Enormously versatile molecules found throughout nature, as well as synthetic compounds with numerous commercial applications. Polymers consist of long chains of atoms, and their physical versatility is derived from variations in the type, number and arrangement of those atoms. The chain is constructed from small repeating units, or monomers. Examples of polymers include lignin (which makes Christmas trees woody), starch in potatoes and the cellulose in paper decorations. Perhaps the most important is DNA, the genetic blueprint for living organisms.

PROTEIN   A class of large molecules found in living organisms, consisting of strings of AMINO ACIDS folded into complex but well-defined three-dimensional structures. The proteins myosin and actin make up the fibres that give turkey muscle its texture.

QUANTUM GRAVITY   A much sought-after theory of every-thing, straddling the world of the very small, as described by quantum theory, and the very big, as described by Einstein's theory of gravity.

QUANTUM THEORY   The most revolutionary scientific theory of the century, applicable to lasers, micro-electronics and Advent candles. It was pioneered between 1900 and 1928, mainly by European physicists who realized that the previous 'classical' theories did not work when applied to subatomic particles, since at this microscopic level, energy changes in abrupt and tiny jumps (quantum leaps). Electrons can be seen making these jumps during firework displays or in candle flames; red corresponds to a small quantum leap and blue to a relatively big one.

REACTION   In chemistry, the coming together of ATOMS or molecules, causing a chemical change to take place that rearranges the atoms.

RECEPTORS   Triggering devices in the membranes of our cells that allow drugs to work. Hormones and drugs are the keys that open the receptors' locks. When the chemical activates a given receptor, it will trigger a specific response. For example, sex hormones activate their receptors to determine whether a fertilized egg develops into a boy or a girl. When we eat too much at Christmas, a locally acting hormone called a prostaglandin sends the message to deposit fat by docking with a receptor.

REDUCTIONISM   A doctrine according to which complex

phenomena can be explained in terms of something simpler. Atomistic reductionism contends that macroscopic phenomena, such as the heady aroma of mulled wine, can be explained in terms of the properties of ATOMS and molecules.

RELATIVITY   Theory dealing with the concepts of space, time and matter, developed by Albert Einstein (1879–1955) as an extension of ideas first set out by Sir Isaac Newton. The effects relativity predicts lie at the boundaries of our experience, in the domains of the supersmall, the superfast, the superlarge and the supermassive. *Special relativity*, announced in 1905, starts from the premise that the speed of light and the laws of physics are the same – invariant – for observers moving at constant speeds relative to one another. *General relativity*, unveiled a decade later, extends this idea, introducing the principle that the laws of physics should be the same for all observers, regardless of how they are moving relative to one another. Conceived after Einstein realized that a man falling off a roof would not feel his own weight, general relativity describes gravity not as a force (as Newton had done) but as the curvature of 'spacetime', a four-dimensional mix of space and time. The theory may help to explain how Santa delivers all those presents on Christmas Eve.

RNA (ribonucleic acid)   The genetic material used to translate DNA into proteins. In some cases, it is also the principal genetic material in a VIRUS.

SECOND LAW OF THERMODYNAMICS   The novelist and scientist C.P. Snow said that this law should be part of the intellectual complement of any well-educated person. Put baldly, it states that a quantity called ENTROPY always increases in an isolated system so that things tend to get more disorganized – a fancy way of saying that molecules in a puddle of water do not spontaneously rearrange to form a snowman.

SEX CHROMOSOME   The Y-chromosome is the essence of human maleness. Inheritance of this chromosome, one of the bundles of DNA found in our cells, is what determines that a human embryo will be a boy. It is paired with the other sex chromosome, the X, which carries far more information. Men are XY and women are XX.

SNOW   Precipitation in the form of small ice crystals, which may fall singly or in tangled masses called flakes. The crystals are formed in clouds from water vapour.

SOLSTICE   One of the two days in the year when the Sun reaches its greatest excursion north (summer solstice) or south (winter solstice) of the Equator.

STATISTICAL MECHANICS   The discipline that attempts to express the properties of macroscopic systems (large enough to be seen with the naked eye) in terms of their atomic and molecular constituents.

SUPERCOMPUTER   The fastest, most powerful type of COMPUTER.

TERA-   A prefix indicating one million million.

THERMODYNAMICS   The science of heat and work.

UPC   The most ubiquitous sequence of black and white stripes on earth. Stands for Universal Product Code, better known as the bar code. Its pattern of bars and spaces can be read by a scanning device connected to a computer and is now widely used by retailers to keep track of stock, from food to books to records, and to help monitor the surge in sales that takes place every Christmas.

VIRTUAL REALITY   Advanced form of COMPUTER simulation in which a participant is plunged into an artificial environment.

VIRUS One of the smallest infectious agents, consisting of a piece of GENETIC CODE wrapped in PROTEIN, measuring between 15 and 300 nanometres across (one nanometre is one thousand millionth of a metre). Responsible for a huge range of diseases, such as influenza and that cold that often seems to strike around Christmas. It is debatable whether viruses are living, since they have to hijack the molecular machinery of our cells to reproduce (they do this by 'reprogramming' our cells with their genetic code, turning them into virus factories). Although safe, effective drugs do exist for some viral infections, it is difficult to combat viruses with drugs without also harming the cells they parasitize. This is why the common cold is so difficult to treat.

Bibliography

ADDINALL, Peter, 'A Response to R.J. Berry on "The Virgin Birth of Christ"', *Science & Christian Belief*, vol. 9, no. 1 (1997), pp65–72

ATKINS, Peter, *Molecules* (New York: Scientific American Library, 1987)

BARHAM, Peter, *The Science of Cooking* (Berlin: Springer-Verlag, 2000)

BELK, Russell, 'It's the Thought That Counts: A Signed Diagraph Analysis of Gift-giving', *Journal of Consumer Research*, vol. 3 (1976), pp155–62

BELK, Russell, 'Gift-giving Behaviour', *Research in Marketing*, vol. 2 (1979), pp95–126

BENTLEY, W.A., and HUMPHREYS, W.J., *Snow Crystals* (New York: Dover, 1962)

BENTLEY, W.A., and PERKINS, G.H., 'A Study of Snow Crystals', *Appleton's Popular Science*, vol. 53 (1898), pp75–82

BERRY, Sam, 'A Response to P. Addinall', *Science & Christian Belief*, vol. 9, no. 1 (1997), pp73–8

BERRY, Sam, 'The Virgin Birth of Christ', *Science & Christian Belief*, vol. 8, no. 2 (1996), pp101–10

BJÖRNTORP, Per, 'Obesity', *The Lancet*, vol. 350 (1997), pp423–6

BLANCHARD, Duncan, 'Wilson Bentley, Pioneer in Snowflake Photo-micrography', *Photographic Applications in Science, Technology and Medicine*, vol. 8, no. 3 (1973), pp26–8, 39–41

BONYTHON, Elizabeth, *King Cole: A Picture Portrait of Sir Henry Cole, KCB, 1808–82* (London: Victoria and Albert Museum)

BOYER, Bryce, 'Christmas Neurosis', *Journal of the American Psychoanalytic Association*, vol. 3, no. 3 (1955), pp467–88

BRAUN, J., GLEBOV, A., GRAHAM, A.P., MENZEL, A., TOENNIES, J. P., 'Structure and Phonons of the Ice Surface', *Physical Review Letters*, vol. 80 (1998), p2,638

270

Bibliography

BULMER-THOMAS, Ivor, 'The Star of Bethlehem', *Quarterly Journal of the Royal Astronomical Society*, vol. 3, no. 4 (1992), pp363–74

BUTTS, Robert, *William Whewell's Theory of Scientific Method* (Pittsburgh: University of Pittsburgh Press, 1968)

CAPLOW, T., 'Christmas Gifts and Kin Network', *American Sociological Review*, vol. 47 (1982), pp383–92

CARSLAW, H.S., and JAEGER, J.C., *Conduction of Heat in Solids* (Oxford: Oxford University Press, 1959)

CHEAL, D., *The Gift Economy* (New York. Routledge, Chapman and Hall, 1988)

CHOWN, Marcus, 'O Invisible Star of Bethlehem', *New Scientist*, vol. 148, no. 2009/2010 (1995), pp34–5

CLAYTON, Chris, 'Bethlehem's Star', *Astronomy Now* (December 1996), pp57–9

DE COURCY, Geraldine, *Christmastide in Germany* (Bonn: Inter Nationes, 1957)

DAWKINS, Richard, *Unweaving the Rainbow* (London: Allen Lane,1998)

DAY, Peter, and CATLOW, Richard, *The Candle Revisited* (Oxford: Oxford University Press, 1994)

DICKENS, Charles, *The Christmas Books*, vol. I (London: Penguin Books, 1985)

DOLARA, P. et al., 'Analgesic Effects of Myrrh', *Nature*, vol. 379, no. 6,560 (1996), p29

EAST, Robert, *Consumer Behaviour: Advances and Applications in Marketing* (Hemel Hempstead: Prentice Hall, 1997)

EMSLEY, John, *The Consumer's Good Chemical Guide* (Oxford: W.H. Freeman, 1994)

EMSLEY, John, *Molecules at an Exhibition* (Oxford: Oxford University Press, 1998)

EPSTEIN, David and Raphael, 'Bentley's Magnificent Obsession', *National Wildlife* (December 1963/January 1964), pp32–4

FARADAY, M., *The Chemical History of a Candle* (London: Chatto & Windus, 1908)

FORD, Brian, 'Even Plants Excrete', *Nature*, vol. 323, no. 6,090 (1986), p763

FURNHAM, Adrian, 'Beware of Relations Bearing Gifts', *New Scientist*, vol. 120, no. 1,644 (1988), p80

FURST, P., 'Mushrooms – Psychedelic Fungi' in *The Encyclopaedia of Psychoactive Drugs* (London: Burke Publishing, 1986)

GOLBY, J.M., and PURDUE, A.W., *The Making of Modern Christmas* (London: B.T. Batsford, 1986)

GOTODA, Takanari, 'Born in Summer?', *Nature*, vol, 377, no. 6,551 (1995), p672

GREENBERG, Leon, 'Alcohol in the Body', *Scientific American*, vol. 189, no. 6 (1953), pp86–90

HAGSTROM, Warren, 'What Is the Meaning of Santa Claus?', *The American Sociologist*, vol. 1 (1966), pp248–54

HALVORSEN, Odd, 'Epidemiology of Reindeer Parasites', *Parasitology Today*, vol. 2, no. 12 (1986), pp334–9

HAPGOOD, Fred, 'When Ice Crystals Fall from the Sky Art Meets Science', *Smithsonian*, vol. 6, no. 10 (1976), pp67–73

HARDING, Patrick, LYON, Tony, and TOMBLIN, Gill, *How to Identify Edible Mushrooms* (London: HarperCollins, 1996)

HARRISON, Albert, Nancy STRUTHERS and Michael MOORE, 'On the Conjunction of National Holidays and Reported Birthdates: One More Path to Reflected Glory?', *Social Psychology Quarterly*, vol. 51, no. 4 (1988), pp365–70

HILLIER, Bevis, *Greetings from Christmas Past* (London: The Herbert Press, 1982)

HOFFMAN, P., KAUFMAN, A., HALVERSON, G., SCHRAG, D., 'A Neoproterozoic Snowball Earth', *Science*, vol. 281 (1998), pp1342–1346

HOLMES, Michael, 'Revolutionary Birthdays', *Nature*, vol. 373, no. 6,514 (1995), p468

HUGHES, David, *The Star of Bethlehem Mystery* (London: Dent, 1979)

HUMPHREYS, Colin, 'The Star of Bethlehem', *Science & Christian Belief*, vol. 5, no. 2 (1993), pp83–101

HUMPHREYS, Colin, and WADDINGTON, W.G., 'Dating the Crucifixion', *Nature*, vol. 306, no. 5945 (1983), pp743–6

KEATHLEY, D.E., 'Biological Enhancement of Christmas Tree Production in Michigan', *Michigan Christmas Tree Journal*, no. 36 (1993), pp38–40

KEVERNE, Eric, MARTEL, Fran, and NEVISON, Claire, 'Primate Brain Evolution: Genetic and Functional Considerations', *Proceedings of the Royal Society London*, vol. 262 (1996), p689

KIDGER, M., *The Star of Bethlehem* (Princeton: Princeton University Press, 1999)

KOENIG, Harold, *Is Religion Good for Your Health?* (New York: Haworth Press, 1997)

KOENIG, H.G., COHEN, H.J., BLAZER, D.G., et al., 'Religious Coping and Depression in Elderly Hospitalized Medically Ill Men', *American Journal of Psychiatry*, vol. 149 (1992), pp1693–1700

KOENIG, H.G., et al., 'Religion and Anxiety Disorder: an examination and comparison of associations in young, middle-aged, and elderly adults', *Journal of Anxiety Disorders*, vol. 7 (1993), pp321–42

KOENIG, H.G., et al., 'Attendance at Religious Services, Interleukin-6, and Other Biological Indicators of Immune Function in Older Adults', *International Journal of Psychiatry in Medicine*, vol. 27 (1997), pp233–50

KOENIG, H.G., GEORGE, L.K., COHEN, H.J., HALES, J.C., BLAZER, D.G., LARSON, D.B., 'The Relationship between Religious Activities and Blood Pressure in Older Adults', *International Journal of Psychiatry in Medicine*, Vol. 28 (1998, pp189–213)

LAROCHE, M., Kim, C., SAAD, G., and BROWNE, E., 'Determinants of In-Store Information Search Strategies Pertaining to a Christmas Gift Purchase' (working paper, Concordia University, 1997)

LEADER-WILLIAMS, N., *Reindeer on South Georgia* (Cambridge: Cambridge University Press, 1988)

MCCULLOUGH, M.E., LARSON, D.B., KOENIG, H.G., MILANO, M.G., 'Systematic Review of Published Research on Religious Commitment and Mortality, 1967–1996', *Journal of the American Medical Association* (1997, in submission)

MCELDUFF, Patrick, and DOBSON, Annette, 'How Much Alcohol and How Often?', *British Medical Journal*, vol. 314 (1997), pp1159–64

MCGEE, Harold, *On Food and Cooking: The Science and Lore of the Kitchen* (New York: Collier Books, 1984)

MARTIN, W.T., 'Religiosity and United States Suicide Rates, 1972–1978', *Journal of Clinical Psychology*, vol. 40, no. 5 (1984), pp1166–9

MATTHEWS, Robert, 'Odd Socks: A Combinatoric Example of Murphy's Law', *Mathematics Today* (March–April 1996), pp39–41

MATTHEWS, Robert, 'Hurry Up and Wait', *New Scientist* (19 July 1997), pp24–7

MILLER, Daniel (ed), *Unwrapping Christmas* (Oxford: Clarendon Press, 1995)

MOLNAR, M.R., 'An Explanation of the Christmas Star Determined from Roman Coins of Antioch', *The Celator*, vol. 5, no. 12 (1991), pp8–12

MOLNAR, M.R., 'The Coins of Antioch', *Sky & Telescope*, vol. 83 (1992), pp37–9

MOLNAR, M.R., 'The Magi's Star from the Perspective of Ancient Astrological Practices', *Quarterly Journal of the Royal Astronomical Society*, vol. 36 (1995), pp109–26

MOLNAR, M.R., *The Star of Bethlehem* (London: Rutgers University Press, 1999)

MONTAGUE, Carl, et al., 'Congenital Leptin Deficiency Is Associated with Severe Early-onset Obesity in Humans', *Nature*, vol. 387, no. 6,636 (1997), pp903–8

MOORE, Peter, 'Why Be an Evergreen?', *Nature*, vol. 173, no. 5,996 (1984), p703

MORRIS, Desmond, *Christmas Watching* (London: Jonathan Cape, 1992)

MULLET, Mary B., 'The Snowflake Man', *American Magazine*, vol. 99 (1925), pp28–31

NITTMANN, J., and STANLEY, H.E., 'The Connection between Tip-splitting Phenomena and Dendritic Growth', *Nature*, vol. 321, no. 4/5 (1986), pp663–8

NITTMANN, J., and STANLEY, H.E., 'Non-deterministic Approach to Anisotropic Growth Patterns with Continuously Tunable Morphology', *Journal of Physics A*, vol. 20, no. 4/5 (1987), p1185

NORTH, Adrian, and HARGREAVES, David, 'The Musical Milieu: Studies of Listening in Everyday Life', *The Psychologist*, vol. 10, no. 7 (1997), pp309–12

OHLSSON, R., HALL, K., and RITZEN, M. (eds), *Genomic Imprinting: Causes and Consequences* (Cambridge: Cambridge University Press, 1995)

OTNES, C., KIM, Y., LOWREY, T., 'Ho, Ho, Woe: Christmas Shopping for "Difficult" People', *Advances in Consumer Research*, vol. 19 (Provo, Utah: Association for Consumer Research, 1992)

OXMAN, T.E., FREEMAN, D.H., MANHEIMER, E.D., 'Lack of social participation or religious strength and comfort as risk factors for death after cardiac surgery in the elderly', *Psychosomatic Medicine*, vol. 57 (1995), pp5–15

Bibliography

PARGAMENT, Kenneth, *The Psychology of Religious Coping: Theory, Research, and Practice* (New York: Guilford Press, 1997)

PARKINSON, Claire, *Breakthroughs: A Chronology of Great Achievements in Science and Mathematics*, 1200–1930 (London: Mansell Publishing Ltd, 1985)

POLLAY, R., 'It's the Thought That Counts: a case study in Christmas excesses', in *Advances in Consumer Research*, vol. 14 (Provo, Utah: Association for Consumer Research, 1986)

POND, C.M., MATTACKS, C.A., COLBY, R.H. and TYLER, N.J., 'The anatomy, chemical composition and maximum glycolytic capacity of adipose tissue in wild Svalbard reindeer (*Rangifer tarandus platyrhynchus*) in winter', *Journal of the Zoological Society of London*, vol. 229 (1993), pp17–40

POND, C.M., *The Fats of Life* (Cambridge: Cambridge University Press, 1998)

POND, C.M., 'An Evolutionary and Functional View of Mammalian Adipose Tissue', *Proceedings of the Nutrition Society*, vol. 51 (1992), pp367–77

POND, C.M., 'The Structure and Organization of Adipose Tissue in Naturally Obese Non-hibernating Mammals', in *Obesity in Europe '93: Proceedings of the Fifth European Congress of Obesity* (London: J. Libbey & Co, 1994), pp419–26

PRETL, George, CUTLER, Winnifred, CHRISTENSEN, Carol, HUGGINS, George, GARCIA, Celso-Ramon and LAWLEY, Henry, 'Human Axillary Extracts', *Journal of Chemical Ecology*, vol. 13, no. 4 (1987), pp717–31

PROEBSTING, Bill, and MONTANO, Jose, 'Needle Abscission in Douglas Fir', *Christmas Tree Lookout*, vol. 23, no. 2 (1990), pp30–32

RIDLEY, Matt, *The Origins of Virtue* (London: Viking, 1996)

RUDGLEY, R., *The Alchemy of Culture* (London: British Museum Press, 1993)

SAMUEL, Delwen, 'Investigation of Ancient Egyptian Baking and Brewing Methods by Correlative Microscopy', *Science*, vol. 273 (1996), pp488–90

SAMUEL, Delwen, 'Archaeology of Ancient Egyptian Beer', *Journal of the American Society of Brewing Chemists*, vol. 54, no. 1 (1996), pp3–12

SANSOM, William, *Christmas* (London: Weidenfeld and Nicolson, 1968)

SCHATZMAN, Morton, 'Does Christmas Drive You Crackers?', *New Scientist*, vol. 120, no. 1644 (1988), pp46–8

SEN, S., et al., 'In-vitro Micropropagation of Afghan Pine', *Canadian Journal of Forest Research*, vol. 24 (1994), pp1248–52

SEN, S., et al., 'Micropropagation of Conifers by Organogenesis', *Plant Physiology*, vol. 12 (1993), pp129–35

SHERRY, J., and MCGRATH, M., 'Unpacking the Holiday Presence: A Comparative Ethnography of Two Gift Stores', in E. Hirschmann (ed) *Interpretive Consumer Research* (Provo, Utah: Association for Consumer Research, 1989)

SIMONS, Paul, *Weird Weather* (London: Little Brown and Company, 1996)

SINGER, Charles, et al., *History of Technology*, vol. 4 (Oxford: Clarendon Press, 1958)

SMITH, Sheryl, et al., 'GABA receptor alpha-4 subunit suppression prevents withdrawal properties of an endogenous steroid', *Nature*, vol. 392, no. 6,679 (1998), pp926–30

SMITH, T.K., MUSK, S.R.R., and JOHNSON, I.T., 'Allyl isothiocyanate selectively kills undifferentiated HT29 cells in vitro and suppresses aberrant crypt foci in the colonic mucosa of rats', *Biochemical Society Transactions*, vol. 24 (1996), p381

STERBA, Richard, 'On Christmas', *Psychoanalytic Quarterly*, no. 13 (1944), pp79–83

STERN, Kathleen, and MCCLINTOCK, Martha, 'Regulation of Ovulation by Human Pheromones', *Nature*, vol. 392, no. 6,672 (1998), p177.

STODDARD, Gloria May, *Snowflake Bentley, Man of Science, Man of God* (St. Louis: Concordia Publishing, 1979)

VERHAGEN, Hans, et al., 'Reduction of Oxidative DNA-damage in Humans by Brussels Sprouts', *Carcinogenesis*, vol. 16, no. 4 (1995), pp969–70

VINES, Gail, 'My Best Friend's a Brussels Sprout', *New Scientist*, vol. 152, no. 2,061 (December 1996), pp46–9

YEO, Richard, *Defining Science: William Whewell, Natural Knowledge, and Public Debate in Victorian Britain* (Cambridge: Cambridge University Press, 1993)

Index

Index